SCIENTIFIC WRITING: A COMPLETE GUIDE

SCIENTIFIC WRITING: A COMPLETE GUIDE

MITCHELL P. JONES PH.D.

ISBN: eBook: 978-0-6451654-5-6
Paperback: 978-0-6451654-9-4

The publisher and the author make no guarantees concerning the level of success you may experience by following the advice and strategies contained in this book. Results will differ for each individual. The content provided in this guide is based on personal experience and anecdotal evidence, is subjective and a matter of author preference. The content may also lend itself to certain scientific fields, types of manuscripts and styles of writing better than others.

Although the author and publisher have made every reasonable attempt to ensure the complete accuracy of the content provided in this guide, they accept no responsibility for errors or omissions. The content is intended to inform the reader and should not replace common sense, legal, medical or other professional advice.

Examples provided in this book relate to the aging behaviour of untreated and diimide treated (hydrogenated) natural rubber balloons at ambient and elevated temperatures. These are only included with the intention of illustrating writing principles in an easily understandable and accessible fashion for readers of all backgrounds. No scientific value whatsoever is attributed to this content. The information provided in the examples is not suggested to be accurate, relevant or suitable research material for a scientific manuscript. It serves only to demonstrate the points made on the writing process.

Any trademarks, service marks, product names, or named features are assumed to be the property of their respective owners and are used only for reference. The author has no affiliation whatsoever with any of the companies or products referenced in this book and no financial or any other incentive is derived from such reference. Reference to a product does not constitute an endorsement.

TABLE OF CONTENTS

WHAT IS SCIENTIFIC WRITING? 1

BEFORE YOU START 3

USING THIS GUIDE 5

Monograph Style Bachelor's, Master's or Doctoral Thesis 6

A. Thesis Literature Review Chapter 8

Bachelor's and Master's Theses 9

Doctoral Theses 10

B. Research Chapter(s) 10

Bachelor's and Master's Theses 10

Doctoral Theses 11

C. Thesis Conclusions and Suggestions for Future Research Chapter 12

D. Thesis Introduction Chapter 13

E. Thesis Abstract 14

F. Cover Page, Declaration and Acknowledgements 15

G. Lists of Thesis Content and Structure 16

Lists of symbols, abbreviations and acronyms 16

Lists of Equations, Figures and Tables 17

Notes on Formatting Equation, Figure and Table Captions in Microsoft Word 17

Table of Contents 17

Notes on Text and Heading Formatting in Microsoft Word .. 18
Notes on Cross-referencing in Microsoft Word 18
Notes on Header and Footer Formatting in Microsoft Word 19
Notes on Page Numbering in Microsoft Word 19
H. List of Publications and Author Contributions Statement 20
I. References 21
J. Appendices 21
Doctoral Thesis by Publication 21
Journal Articles 23
Common Journal Article Types 24
Journal Reputation and Metrics 25
Journal Article Access and Licensing 26
Supplementary Material 27
Conference Papers 27
Scientific Programme Structure 28
Conference Presentation Format 28
Conference Costs 29
Opportunities for Conference Delegates 29
Conference Publications 30
Books and Book Chapters 30
Book and Book Chapter Costs 31
Book and Book Chapter Authorship 32
Publisher-Contributor Agreements 32
Writing and Submitting a Book or Book Chapter 33
Indexing and Author Profiles 33

Citations 34
Self-citation 34
Author Profiles 35
Open Researcher and Contributor ID 35
WRITING AN ORIGINAL RESEARCH MANUSCRIPT 37
Writing Order 37
Outline of Results and Discussion 38
Figure and Table Design 44
Micrographs 44
Magnification and Scale 45
Brightness and Contrast 46
Cropping Micrographs 47
Adding a Scale Bar 48
Graphs 50
Graph Selection 51
Grouping Graph Content 52
Graph Layout Basics 53
Using Multiple Axes 56
Line and Point Formatting 57
Representing Error 59
Labelling and Legends 60
Improving Data Layout 61
Highlighting the Message 62
Exporting Graphs 63
Tables 65

Table Formatting 65
Number Formatting and Appropriate Use of Decimal Places 68
Schematics 69
Generating and Combining Picture Elements 69
Generating and Styling Shapes 71
Figure and Table Captions 73
Unit Formatting 76
Equations 77
Results and Discussion 79
Internal Comparison 87
Describing Data 88
Improving Communication of Data 90
External Comparison and Discussion 91
Scientific Writing Tips 94
If something is worth saying – then make it black and white 94
Be compulsively concise and precise 96
Conclusion 99
Central Message 100
Key Results 101
Manuscript Value and Impact 103
Author Comment and Future Perspective 104
Materials and Methods 106
Allocating Content to the Materials and Methods 106
Materials 107
Sample Preparation 108

Experimental and Analytical Methods 111
Introduction 114
Final Paragraph – Aims and Execution 114
Preceding Paragraphs 115
Topic and Relevance 116
Motivation and Research Gap 118
Selecting References for the Introduction 121
Abstract 122
Topic and Relevance 124
Motivation and Research Gap 126
Aims and Execution 127
Key Results 129
Manuscript Value and Impact 130
Selecting a Title 132
Keywords 134
References 136
Reference Management Software 137
Using Exportable Citations 138
Using Reference Style Templates 139
Converting to Plain Text 140
LITERATURE REVIEW MANUSCRIPTS 143
Improving Review Article Impact in Journals 143
Writing a Literature Review 145
Reusing Published Content in a Review 150
REVIEW AND EDITING 153

Reviewing Theses Prior to Submission 153

Reviewing Publications with Multiple Authors Prior to Submission 155

Author Revisions in Series 155

Author Revisions in Parallel 156

SUBMITTING A THESIS 159

Bachelor's Programmes 159

Master's and Doctoral Programmes 160

Thesis Examination 160

Student Revisions 162

Response to Examiners' Comments 162

Revised Thesis 163

Public Thesis Defence 164

Thesis Archiving 165

PUBLISHING JOURNAL ARTICLES 167

Selecting a Journal 167

Authorship 168

First Author 170

Co-authors 170

Last Author 171

Corresponding Author 171

Affiliations 172

Formatting a Manuscript for Submission 174

Using the Guide for Authors 174

Understanding Journal Expectations 174

Journal Abstracting and Indexing 175
Getting to Know the Journal Editors 176
Flexibility in Formatting of Initial Submissions 176
Traditional Page Formatting Requirements 177
Content Requirements 178
Electronic Artwork Requirements 179
Data Availability Requirements 179
Less Traditional Content Requirements 181
Common Submission Supplements 182
Graphical Abstract 182
Highlights 184
Cover Letter 185
Supplementary Material 186
Manuscript Submission 186
Manuscript Information 188
File Upload and Submission 190
Peer Review and Revisions 192
Desk Rejection by the Editor 192
The Peer Review Process 193
Reviewer Assignment 193
Peer Review Timelines 194
Type and Scope of Reviewer Feedback 194
Editorial Decision – Acceptance, Rejection or Revision 195
Manuscript Revisions and Resubmission 196
Response to Reviewers' Comments 196

Revised Manuscript 197
Resubmission Cover Letter 198
Resubmission 198
Article Production and Correction of Proofs 199
Final Paperwork 200
AUTHOR PROFILE 201

WHAT IS SCIENTIFIC WRITING?

Scientific writing does not simply refer to text which contains scientific content. There are actually many styles of writing which can incorporate scientific themes. Newspapers, for example, often refer to science but do not use a scientific writing style. Scientific writing instead refers to the style of writing used in scientific texts, such as theses, journal articles, books and conference documents, to communicate new ideas and findings.

Emphasis in scientific writing is placed on clarity and precision. There is no room for poetic or creative value. It must be concise, to-the-point and accessible to all audiences. Short factual sentences omitting word play and unnecessary content are common. Since scientific writing is typically used to prove facts and disprove false hypotheses, word selection and use is very important. It is critical that authors are very clear and unambiguous with regard to what exactly they claim. The writing should also include as little bias as possible. Use of persuasive or emotive language is not common.

Simple, clear and unopinionated expression is particularly important given the international nature of the scientific community and hence readership. English may not be the first language of many readers. Even if it is, scientific research is so multidisciplinary that readers may not be familiar with much of the content and related terminology within a scientific text. Jargon is subsequently avoided where possible, although this can sometimes be at odds with maintaining brevity and precision. The degree of tolerance for jargon tends to depend on the document type and intended purpose. Some highly influential journals with broad scopes are critical of jargon but in most cases its use is begrudgingly accepted.

Perhaps most importantly, scientific writing does not have the freedom of style and content that other writing might enjoy. There are expectations on how a manuscript will be formatted and what types of content it will contain. An unofficial international standard exists which ensures that, regardless of which of the millions of scientific documents published every year is opened, the reader knows what types of content it will contain and where to find them. The intricacies of the presentation of this content are the focus of this guide.

BEFORE YOU START

This book covers the scientific writing process. Its scope does not extend to research conceptualisation, design of experiment and experimental work. These preliminary phases should be completed with an experienced supervisor prior to writing and typically encompass:

- Literature review to establish what is known on the topic, the current challenges, unknowns, and opportunities for improvement.
- Identification of a research gap i.e., something relating to the topic that has not yet been reported in the literature.
- Definition of research aims i.e., what your research hopes to achieve.
- Division of these aims into research questions that are focussed and specific enough to be comprehensively and definitively answerable within the project duration.
- Construction of a test matrix defining the materials, methods, test conditions, variables, instruments and equipment that will form the basis of the experimental work.
- Experimental work based on this test matrix.

Although the content of this guide is also generically useful it is best suited for use with well-planned and structured research completed based on these preliminary steps. The scientific writing phase should ideally only begin once these steps have been finalised.

USING THIS GUIDE

The general consensus is that scientific writing is too broad a subject to permit the provision of comprehensive and useful advice using generic instructions. While this may be true, the exceptionally vague tips often offered to students on the subject can be far from helpful. This book aims to provide a structured process with an instructional tone to guide those new to scientific writing through the production of their scientific manuscript, be it a bachelor's, master's, or doctoral thesis, journal article, conference paper or book chapter. This complete guide covers all of these manuscript types and can subsequently be hard to follow for beginners. It is intended for the intermediate reader seeking an overview of all content. Beginners are directed to the specifically formatted sub guides available for each document type. Most aspects of this guide are relevant to all scientific documents but specific information on how best to use the guide for each manuscript type is also provided.

Some topics covered in this book and some recommendations provided are subjective and are a matter of personal preference. The content may also lend itself to certain scientific fields, types of manuscripts and styles of writing better than others. Scientific writing, like any writing style, is to some degree open to interpretation and does vary by field, community, and culture. The individual requirements of each manuscript type and journal content and stylistic regulations will also govern suitable manuscript content and formatting. The suggestions provided in this guide are intended only as ideas of good practice in scientific writing.

Examples provided in this book relate to the aging behaviour of untreated and diimide treated (hydrogenated) natural rubber balloons at ambient and elevated temperatures. These are only included with the intention of illustrating writing principles in an easily understandable and accessible fashion for readers of all backgrounds. No scientific value whatsoever is attributed to this content. Hydrogenation and thermal aging tests are most relevant to elastomers exposed to harsh service conditions e.g., automotive components. Balloons were selected as a simpler, more relatable and less engineering-specific hypothetical example application. The conclusions are also hypothetical: increasing the shelf life of balloons to reduce waste is at odds with improving their biodegradability, an arguably more important property given the short service life of balloons. The information provided in the examples in this guide is subsequently not suggested to be accurate, relevant or suitable research material for a scientific manuscript. It serves only to demonstrate the points made on the writing process.

Most software specific instructions refer to Microsoft® Word (Microsoft 365®), the Fiji distribution of ImageJ (freeware) and OriginLab OriginPro® 2022.

As a general tip, regardless of what sort of scientific manuscript is being produced, work using document versions. Save considerable progress or revisions as new files and label them with version numbers. This can be useful if a document becomes lost or corrupted and allows self-regulated version histories beyond automatic features provided in Microsoft Word. Back all files up regularly to an online location or external storage device to ensure that no work is lost if a problem occurs with an electronic file or computer.

MONOGRAPH STYLE BACHELOR'S, MASTER'S OR DOCTORAL THESIS

Completing a thesis is usually a prerequisite requirement for the award of a master's, or doctoral degree at a university. Bachelor's degrees in many countries also incorporate some form of thesis requirement. The standard thesis comprising a unified text broken into appropriately titled chapters is considered a 'monograph' style thesis.

Each university administration usually has their own template, requirements and expectations for thesis preparation, which can vary widely even between supervisors. A comprehensive understanding of the applicable thesis requirements is subsequently critical before proceeding.

Thesis components are listed in TABLE 1 in the order in which they typically appear in a thesis. However, this guide will discuss them in the order in which they should be considered and prepared as denoted by steps A-J. Students should follow these steps to prepare all necessary sections and then compile them in the correct order at the end.

TABLE 1. *Overview of the components typically presented in bachelor's, master's and doctoral theses and their variation by degree type. Table content is listed in the order in which it usually appears in a thesis. The most logical writing sequence is denoted using steps A-J.*

Step	*Thesis Component (order may vary)*	*Bachelor*	*Master*	*Doctorate*
F	Cover page	×	×	×
	Declaration	×	×	×
	Acknowledgements	×	×	×
E	Abstract	×	×	×
H	List of publications			×
	Author contributions statement			×
G	List of symbols	×	×	×
	List of acronyms	×	×	×
	List of abbreviations	×	×	×
	List of equations	×	×	×
	List of figures	×	×	×
	List of tables	×	×	×
	Table of contents	×	×	×
D	Thesis introduction chapter		×	×
A	Thesis literature review chapter		×	×
B	Thesis materials and methods chapter (option 1)		×	×
B	*Research chapter(s)*	1	1+	3-5+
	Chapter introduction	×	×	×
	Integrated literature review	×		
	Chapter materials and methods (option 2)	×	×	×
	Chapter results and discussion	×	×	×

Step	Thesis Component (order may vary)	Bachelor	Master	Doctorate
	Chapter conclusion	×	×	×
C	Thesis conclusions and suggestions for future research chapter		×	×
I	References	×	×	×
J	Appendices	×	×	×

A. Thesis Literature Review Chapter

Familiarising yourself with literature e.g., original research and review journal articles, book chapters or books, conference papers and general information available online about your research topic is always the first stage of any research project and should have been completed during the research planning phase, which is outside the scope of this guide and should be completed with an experienced supervisor. During this phase research gaps were identified and research objectives, specific well-scoped and achievable research questions, and a structured methodology defined. This planning enabled the completion of the experimental work that will in turn be described in the thesis.

Bearing that in mind, literature review is an ongoing process. New research is always framed in the context of the state of the literature at the time of writing. A tiny cog in an enormous machine, a scientific thesis should have some timely purpose and relevance to the development of a scientific field. The research work that has been completed was designed based on the state of the literature at the time that it was conceived. Depending on how long the experimental work took, the state of the literature may have changed by the time the thesis can be written. Active research fields move very quickly and in as little as a few months there can be radical advances. It is subsequently a good idea for students to incorporate ongoing literature review practices in their work schedule for the entire duration of their project. At a minimum, the thesis 'Literature Review' content and references must reflect the current state of the literature at the time of submission.

The first step before beginning to document your work is subsequently literature review. A literature review provides an extensive synthesis of an established research field, summarising the results and developments of many existing scientific documents in a single section. It acts as an independent introductory reference for those unfamiliar with a research area and is subsequently an important element in most theses, where a reader must first acquaint themselves with a research field before being able to understand and appreciate the value of a new contribution. Check for new publications in the research area using relevant keywords and an indexing service such as Google Scholar (www.scholar.google.com). What exactly you include in your literature review is up to you, but it should cover all relevant fundamental and background information on the research topic, a comprehensive overview of the existing research in the field and describe the current state of the literature.

Literature reviews, while important, play only a supporting role in a thesis, providing context for the main research work. There is subsequently a fair degree of tolerance from examiners when assessing thesis literature reviews. As long as a thesis literature review is up to date and comprehensively details the current state of the literature for the research area, it is generally considered to have fulfilled its function.

Having already planned and completed the project's experimental work it is assumed that the student is familiar with the research area and its literature. Write or update the literature review for the thesis now using the suggestions provided in the LITERATURE REVIEW MANUSCRIPTS section of this guide. Doing so will also provide a strong background for writing the results and discussion content of the thesis, which must be framed in the context of the current state of the literature and refer to recent developments in the research field.

Bachelor's and Master's Theses

A fairly short literature review encompassing the synthesis of 50-100 references may be sufficient for bachelor's and master's theses. The expected requirements and format of a literature review in a bachelor's or master's thesis vary much more radically by institution than those of a doctoral thesis for which there is a more commonly recognised standard (although even doctoral thesis structure can vary considerably between and within institutions). For example, some universities opt for an integrated literature review in a bachelor's thesis, barely encompassing more than a few paragraphs embedded in or directly following the introduction or other introductory material. At other universities and especially for master's degrees the literature review must be much more extensive and form its own chapter as in a doctoral thesis. The suggestions on writing a literature review provided in this guide relate to this latter option. Be sure to check the university requirements for the literature review before starting. Such a comprehensive review may be deemed unnecessary at your institution.

Doctoral Theses

The literature review in a doctoral thesis is presented as a chapter broken into themed and appropriately titled sections and is typically quite long (6,000-12,000 words, although the appropriate length will vary based on the amount of content published in the research area and the scope of the review). The number of references in the literature review chapter will also vary based on the field and topic, with some research areas encompassing a lot more literature than others. Nevertheless, the number of references should be substantial with several hundred references typically used.

B. Research Chapter(s)

The most important aspect of any thesis is the research chapter(s) that it contains, i.e., the chapters that detail the results and discussion of the experimental work that was completed over the course of the project. This is the core scientific work that is being assessed. Regardless of which type of thesis is being written, the process for writing a research chapter is more or less the same.

Follow the instructions in the WRITING AN ORIGINAL RESEARCH MANUSCRIPT section of this guide to assist you in writing the number of thesis research chapters required for your degree. Any information that is only relevant to another manuscript type, such as a journal article or conference paper, can be ignored. Once the required research chapters have been completed you can return here and continue.

Bachelor's and Master's Theses

Requirements vary by institution; however, most bachelor's theses can be assembled from what would be one single research chapter in a doctoral thesis or a journal article. That means that the steps detailed in the WRITING AN ORIGINAL RESEARCH MANUSCRIPT section of this guide are akin to writing the entire bachelor's thesis. The 'Introduction', 'Materials and Methods', 'Results and Discussion' and 'Conclusion' detailed in this section are simply broken into thesis sections or chapters as appropriate based on the university's expectations and the 'Abstract' used as the thesis abstract. As already discussed, an integrated or chapter-based literature review is also typically included. Once this content has been finalised the thesis is effectively finished.

Master's theses also usually contain a single (or sometimes multiple) research chapters. Universities may adopt an integrated master's thesis structure closer to a bachelor's thesis or utilise the additional extended chapters typical of a doctoral thesis: extended thesis 'Introduction', 'Literature Review', and 'Conclusions and Suggestions for Future Research' chapters. Be sure to familiarise yourself with university expectations on the inclusion or omission of these chapters before continuing. If these additional chapters are required then continue to the sections below.

Doctoral Theses

Doctoral theses typically contain at least two and normally three to five research chapters. While a thesis research chapter can be compared to a journal article, a thesis chapter is more flexible in its layout and content. It can be much longer; it can have a lot more results and discussion; and, in particular, it can include content that may not be deemed important enough to warrant publication in a journal.

A doctoral thesis also typically includes extended thesis 'Introduction', 'Literature Review', and 'Conclusions and Suggestions for Future Research' chapters. Materials and methods may be grouped and presented as an extended chapter detailing all experimental work completed over the course of the thesis in one place (option 1) or as subsections in each research chapter documenting only the experimental work completed in the chapter in question (option 2).

There is no rule on how to present the materials and methods, but it generally makes sense to group them into a single separate thesis chapter when many similar materials and methods are used across all research chapters. This helps to avoid repeating the same information in each chapter. If the materials and methods vary considerably from chapter to chapter, then it makes more sense to list the relevant materials and methods information in each chapter separately. Tips on writing materials and methods can be found in the WRITING AN ORIGINAL RESEARCH MANUSCRIPT — MATERIALS AND METHODS section of this guide.

C. Thesis Conclusions and Suggestions for Future Research Chapter

Bachelor's theses usually contain a short conclusion as opposed to the more extended thesis 'Conclusions and Suggestions for Future Research' chapter of a master's or doctoral thesis. The content in this section of this guide is subsequently unlikely to be relevant for those writing a bachelor's thesis. Instructions on writing conclusions can be found in the WRITING AN ORIGINAL RESEARCH MANUSCRIPT — CONCLUSION section of this guide. For more advanced theses, the extensive dedicated 'Conclusions and Suggestions for Future Research' chapter constitutes a series of paragraphs detailing the conclusions of each thesis research chapter and insight into the future potential of the research area.

Paste the conclusions of each individual research chapter into a new Microsoft Word document. Formulate the 'Conclusions and Suggestions for Future Research' chapter text using these conclusions as inspiration for new free-standing text that flows from one set of ideas and conclusions (thesis chapter) to the next. While this text should only focus on the central messages and key results of the chapters coupled with their impact (already the focus of the pasted conclusions), the logical sequence of presentation should hint at the progressive and interlinked or complementary nature of the steps required to complete the project i.e., a progression of ideas built off each other to achieve a central project goal. Provided the project was well planned and the chapter conclusions well written this is a straightforward process.

Following the conclusions, the student should outline their suggestions for future research. If you were to keep working in the research area, what would you focus on? What would your next steps be? Which concepts have potential? Which don't? What would your recommendation to other researchers starting out in the field be?

Suggestions for future research could potentially relate to:

- Hypotheses.
- New areas of focus.
- Possible relationships that could be investigated.
- Novel or extended applications for the research.
- Experiments worth completing.
- Modifications to existing design of experiment.
- Possible benefits or challenges of new approaches.

D. Thesis Introduction Chapter

Bachelor's theses typically include a relatively short introduction with literature review potentially integrated instead of the extended thesis 'Introduction' chapter of a master's or doctoral thesis. The content in this section of this guide is subsequently unlikely to be relevant for those writing a bachelor's thesis. Instructions on writing introductions can be found in the WRITING AN ORIGINAL RESEARCH MANUSCRIPT — INTRODUCTION section of this guide. For more advanced theses, the extensive dedicated 'Introduction' chapter serves the same purpose as the introductions that would be written for its individual thesis chapters. The main difference is the scope. The introduction of a thesis chapter introduces the topic of the chapter and its relevance, the current state of the literature on the topic, the motivation for the study and how it will be performed. The introductory chapter of the thesis does the same but, rather than focussing on the specific topic and study in a given chapter, it more broadly introduces the project and its associated research area.

Use the instructions on 'topic and relevance' and 'motivation and research gap' in the INTRODUCTION section of this guide to outline the thesis (rather than chapter) topic and relevance, comment on the current state of the literature and the motivation for the work. Being a chapter, this text is typically broken into themed individually numbered sections e.g., '1.1 A Need for Sustainability', '1.2 Consumer Behaviour and Balloon Waste' rather than left as a single piece of text as would be typical of a chapter introduction. This text is intended to provide broad context and background information for the thesis project and can be quite short e.g., a few numbered sections across a few pages. Be concise; provide an overview and ensure that it is strictly scoped to 'topic and relevance' and 'motivation and research gap' related content. Leave excess details for the more comprehensive thesis 'Literature Review' chapter.

The 'aims and execution' aspect of the thesis introduction chapter is more regimented than the final paragraph in chapter introductions and is arranged into the following sections, which should also be numbered:

- Thesis objectives, scope, and research questions which can be concisely presented using bullet points. Limited additional explanation is required beyond the direct expression of this information.
- Thesis methodology, providing a short and broad overview of how the project was completed in a few paragraphs. Paste the final paragraph of the introduction of each chapter (see WRITING AN ORIGINAL RESEARCH MANUSCRIPT — INTRODUCTION — FINAL PARAGRAPH – AIMS AND EXECUTION) into a Microsoft Word document and use them to build a broad overview of the development of the research project over the course of the thesis, supplemented with additional details from the 'Materials and Methods' section of each chapter, as necessary.
- Thesis outline, which simply provides a breakdown of each chapter and a short summary of its contents. Paste the abstract of each research chapter below its heading in the breakdown and use it as a guide for writing each chapter summary. Remember the chapter summary should describe the contents of each chapter in the context of the thesis rather than as a standalone summary (abstract). There should be an obvious logical progression from one chapter summary to the next. Any background information contained in the chapter abstract that relates to the project but has already been described earlier in the thesis 'introduction' chapter should be omitted.

E. Thesis Abstract

While bachelor's theses are so short that the abstract produced in the WRITING AN ORIGINAL RESEARCH MANUSCRIPT — ABSTRACT section of this guide can serve as the thesis abstract, master's and doctoral theses typically require a separate thesis abstract that reflects the full breadth of their content across a potentially wider scope. To create a thesis abstract for these more advanced theses, paste the abstracts associated with each research chapter into a Microsoft Word document in the order that they appear in the thesis. Use this content as a guide for writing an abstract that concisely summarises the entire thesis project rather than a given study presented in one of the thesis chapters. Note that the compiled abstracts should serve only as inspiration for new original text and should not simply be edited to form a new abstract.

The structure and flow of the final thesis abstract should resemble that described in the ABSTRACT section of this guide.

Use the chapter abstracts and thesis 'Introduction' chapter as inspiration for the 'topic and relevance' and 'motivation and research gap' content of the thesis abstract. This content is duplicated throughout the compiled chapter abstracts due to the free-standing nature of each abstract. Include this information only once at the beginning of the thesis abstract as a common abstract section covering the entire thesis.

'Aims and execution' content in the thesis abstract may be presented next as a grouped summary detailing one progressive experimental approach across the course of the project followed by grouped 'key results' content distinguished by chapter. Alternatively, 'aims and execution' content distinct to a chapter's 'key results' can be linked to these results and presented together before moving further to do the same for the next chapter.

The thesis abstract should conclude with a single thesis value and impact abstract section covering all the most important points covered in the 'Conclusions and Suggestions for Future Research' chapter of the thesis. Be sure to keep this broad and applicable to the entire thesis project.

F. Cover Page, Declaration and Acknowledgements

Always check if there is a university template available for formatting thesis submissions. If one exists, it is also worth making sure that it is the most recent version of the template. Many universities will only accept theses that are formatted to their preferred standards. Thesis requirements vary but typically a university template will contain the following sections, which should be completed by the student:

- A cover page listing the thesis title, degree for which the thesis is required, student's name, academic institution, location, and date. Add these basic details.
- A declaration signed by the student to certify that they personally completed the scientific work contained within the thesis, that the thesis has not been submitted to qualify for any other scientific award, that any third-party assistance has been acknowledged and that any relevant ethical guidelines have been adhered to. Read the declaration carefully and then sign it.
- An acknowledgements page detailing any assistance or support that the student received over the course of their project. Completing this page is optional, but any third-party assistance received over the course of the thesis project, such as advice, help with practical work, thesis review or editing, must be acknowledged. If content appears in the thesis and is not attributed to someone else on this page, it is assumed that the student completed the work. If this is not true, then this is an infringement of the declaration that was signed in the previous step.

G. Lists of Thesis Content and Structure

Every thesis typically contains structured lists, which help to organise content and enable the reader to quickly navigate to items that attract their interest. These lists usually appear before the main thesis body and include the following:

Lists of symbols, abbreviations and acronyms

Symbols, abbreviations, and acronyms that are used in the thesis and their associated meaning each warrant their own unique list. When generating these lists read through the thesis carefully and ensure that all terms are accounted for. Arrange symbols, abbreviations, and acronyms in alphabetical order, provide a description for each term and its unit where appropriate.

Lists of Equations, Figures and Tables

The designation, title or description and associated page number of all equations, figures and tables that appear in the thesis are arranged into unique lists in the pages preceding the main thesis content.

Notes on Formatting Equation, Figure and Table Captions in Microsoft Word

Use the Microsoft Word caption tool for all captions appearing in the thesis body to enable automatic generation of the lists of equations, figures and tables that should appear at the start of the thesis. When inserting an equation, figure or table do not type the caption directly into the document. Instead select the space adjacent, under or above the equation, figure, or table as appropriate and choose 'Insert Caption' from the 'References' tab.

Be sure to specify if the caption belongs to an equation, figure or a table in the 'Label' dropdown field. Write the caption and then select 'OK'. After a caption has been inserted, its formatting can be changed e.g., font, size, colour to match the rest of the document.

Once all captions have been completed, a list of equations, figures or tables can be generated by simply selecting 'Insert Table of Figures' from the 'References' tab and then either 'Equation' 'Figure' or 'Table' from the 'Caption Label' dropdown field as appropriate followed by 'OK'.

Table of Contents

The structure and hierarchy of thesis sections should be presented in a table of contents. This can be automatically generated in Microsoft Word by selecting 'Table of Contents' from the 'References' tab and choosing a pre-set from the list. Custom formats can also be defined.

The automatic generation of the table of contents will only work correctly if the Microsoft Word document is formatted properly. Consider the following notes to ensure optimal document formatting for your thesis.

Notes on Text and Heading Formatting in Microsoft Word

Text is typically neatly arranged into paragraphs with an appropriate font, size and spacing (e.g., Arial or Times New Roman in size 11 or 12 pt. with 1.5 pt. spacing and justified alignment). Manuscript text should be set as 'Normal' in the 'Styles' box on the 'Home' tab.

Headings should be arranged hierarchically and formatted using the heading styles in the 'Styles' box on the 'Home' tab of Microsoft Word. If headings are not assigned the correct heading style, then the automatically generated table of contents will in turn not format them correctly. Choose 'Heading 1' for chapter headings, 'Heading 2' for section headings and so on down the hierarchy. Default text formatting e.g., fonts, colour, alignment, sizing, spacing, numbering can be changed for each pre-set style by right clicking on the desired style in the 'Styles' box and selecting 'Modify'. Once the desired formatting has been set all headings that use the corresponding style pre-set in the Word document will be automatically updated to match.

Notes on Cross-referencing in Microsoft Word

Use the cross-reference tool to refer to headings, equations, figures, or tables that appear elsewhere in the manuscript. To do so simply write the main thesis body text as normal and, if at any point the need arises for a reference to another part of the manuscript e.g., another section, an equation, a figure or table, then click 'Cross-reference' on the 'References' tab followed by 'Heading' 'Equation', 'Figure' or 'Table' from the 'Reference type' field. Choose the relevant reference from the list followed by 'Insert'.

Doing so means that, if at some point in the future the labelling of one of these elements changes, then all references to the content in the document can be automatically updated to match. If it is necessary to update these fields, simply right click on the field in question and click 'Update Field' or select all text in the document with the keyboard shortcut Ctrl-A and then use F9 to update the text body.

Notes on Header and Footer Formatting in Microsoft Word

Headers and footers are often used to indicate where pages are located in the thesis structure. Chapter numbers are often listed in the header, chapter titles or descriptions in the header or footer and page numbering in the footer.

Ensure that all chapters are separated by section breaks by selecting the last page of each chapter and opening the 'Layout' tab. Select 'Breaks' and then 'Next Page' from the 'Section Breaks' category of the dropdown menu.

Unlink each header and footer from its previous section in the document by double clicking it and unselecting 'Link to Previous' from the 'Header & Footer' tab. This is especially important if each group of headers and footers should have unique text. 'Different first page' can also be selected from the same tab if the first page of each chapter should have a different header or footer e.g., if the chapter name and number are already written on the first page it may not be necessary to have these details in the header and footer on that page.

Notes on Page Numbering in Microsoft Word

Page numbers are not normally used on the title, dedication, or declaration pages of a thesis. However, they usually appear on all other pages. Roman numerals are sometimes used for the content preceding the main thesis body e.g., acknowledgements, abstract, table of contents, list of publications, author contributions, lists of figures, tables, abbreviations, and acronyms. Numbers are used in the main thesis body.

If the use of different page number styles in various portions of the thesis is desired, then the document must first be broken into sections. Move to the last page of the section that will contain the first style of page numbers (e.g., roman numerals) and open the 'Layout' tab. Select 'Breaks' and then 'Next Page' from the 'Section Breaks' category of the dropdown menu.

Now double click the footer of the first section of the document and add page numbers by selecting the appropriate position and alignment from the 'Page Number' dropdown on the 'Header & Footer' tab. Typically thesis page numbers should appear at the bottom of the page with a right alignment. Return to this menu and select 'Format Page Numbers' to select roman numerals from the 'Number format' dropdown menu. If the document is sectioned correctly only the first section will contain roman numerals.

Move to the next section of the document which should still contain numbers following on sequentially from the roman numerals. Double click the footer of this document section and select 'Format Page Numbers' from the 'Page Number' dropdown on the 'Header & Footer' tab. Enter '1' in the 'Start at' field under 'Page numbering'.

If at any point the sectioning in the document becomes confusing, toggle 'paragraph marks and other hidden formatting symbols' on from the 'Home' tab to make the section breaks visible.

H. List of Publications and Author Contributions Statement

Bachelor's and master's degrees are often too short to generate much publication activity. Conference attendance and the occasional journal article resulting from these degrees are of course possible but there is generally no expectation that bachelor's or master's students will publish over the course of their programme. The content in this section of this guide is subsequently unlikely to be relevant for those writing a bachelor's or master's thesis. Conversely, active participation in scientific dialogue is a common expectation in doctoral programmes, which essentially aim to train students to become effective members of the scientific community. Details of publication activity and conference attendance are highly valued in these degrees and can help to prove that a student has successfully met the requirements of their programme.

Provide a list of all publication activity resulting from your doctoral (or in some cases bachelor's or master's degree) studies prior to the main thesis content, including journal articles, oral or poster presentations at conferences and book chapters. These could be broken into groups based on publication type and listed as Harvard style references in reverse chronological order although the formatting of this list is entirely at the discretion of the author.

A separate list should also be provided detailing the student's contribution to each publication. This is important because, although students can qualify for co-authorship on a publication in many ways, the role that the student performed in the publication must be considerable if that authorship is to be considered a meaningful contribution to the scientific training completed over the course of a doctoral programme. More information on author contributions is provided in the PUBLISHING JOURNAL ARTICLES — AUTHORSHIP section of this guide.

I. References

A reference list detailing all literature used over the course of the entire thesis should be provided at the end of the document in a style that meets the expectations of the university and department to which it will be submitted. Instructions on referencing are provided in the WRITING AN ORIGINAL RESEARCH MANUSCRIPT — REFERENCES section of this guide.

J. Appendices

Appendices are any additional material that is too detailed or insufficiently refined to be useful to the reader. Such content may subsequently be distracting or inappropriate if used in the research chapters but may potentially be important to include in the thesis, nonetheless. Examples could include supporting raw data, calculations, photographs, micrographs, maps, and programming code. This material requires limited explanation and can be directly presented in an appropriately titled appendix e.g., Appendix A – Raw Data for Balloon Material Characterisation. Appendices should be referred to in the manuscript text where appropriate.

Doctoral Thesis by Publication

A doctoral thesis by publication is identical to a standard monograph style doctoral thesis with the exception that its research chapters are replaced by published journal articles or article drafts that have been submitted to but not yet accepted by a journal. Requirements vary by university but typically at least two original research journal articles must be published in reputable journals and at least one further manuscript submitted to a journal at the time of thesis submission for this option to be possible.

The key benefits of submitting a doctoral thesis by publication are reduced thesis preparation time since the published journal articles replace the research chapters, reduced examination time due to the already peer reviewed and published nature of the thesis content and improved impact since journal articles tend to enjoy more exposure and interest than theses in the scientific community.

While it is true that doctoral thesis by publication submissions may receive less scrutiny from examiners than traditional monographs due to the already reviewed nature of the work, they are also likely to be assessed differently, albeit inadvertently. Examiners are well within their rights to expect that the content of any draft chapters that have been submitted with the thesis but not yet accepted by the journal will be published or are at least publishable. This can potentially lead to the research contained within the thesis being scrutinised based on what the examiner considers to be journal editorial standards, rather than the potentially more relaxed thesis examination standards.

Nonetheless, there is fundamentally little structural difference between a traditional monograph style doctoral thesis and a doctoral thesis by publication. Formatting varies by university but one simple method to adopt when writing a thesis by publication is to replace the research chapters with a short 10–20-page summary forming a single chapter entitled 'Summary of Publications: Results and Discussion' or similar, which is broken into sections based on appended journal articles and outlines the key findings of each article and how they relate to each other. This demonstrated continuity is important as it distinguishes the project-based progression of a doctorate from the potentially more random publishing behaviour of established academics. This summary can appear in the position traditionally occupied by the research chapters.

With the articles summarised in this chapter, the published journal articles and submitted drafts can then be included at the end of the thesis as appendices. Reviewers' comments on published articles, supplementary material, and any other article-relevant documents may also be included as appendices. The appended articles and their summary replace only the research chapters of the thesis. All other thesis chapters must still be included and appear in their conventional order. However, since original research journal articles contain a 'Materials and Methods' section by default there is no need for a designated thesis chapter for this content regardless of how similar or different the materials and methods adopted in each study may be.

The structure of a doctoral thesis by publication could resemble the following:

- Cover page.
- Declaration.
- Acknowledgements.
- Thesis abstract.
- List of publications.
- Author contributions statement.
- List of symbols.
- List of acronyms.
- List of abbreviations.
- List of equations.
- List of figures.
- List of tables.
- Table of contents.
- Thesis 'introduction' chapter.
- Thesis 'literature review' chapter.
- 'Summary of publications: results and discussion' chapter.
- Thesis 'conclusions and suggestions for future research' chapter.
- References.
- Appended journal article 1.
- Appended journal article 2.
- Appended journal article draft.

JOURNAL ARTICLES

A scientific journal is basically a themed newspaper or magazine for researchers. It comprises collections of articles arranged by volume and issue. An issue is a set of articles that is published as a single entity at regular intervals e.g., monthly, while a volume typically comprises the issues published over the course of a year. While these collections of articles once represented physical texts, most journals currently maintain almost fully digital operations, publishing content online with hard copies generally only printed to order.

Journal articles are arguably the highest impact and most reputable form of scientific communication. This stems from the often highly selective nature of journal editors when assessing submissions and the rigorous peer review process that manuscripts under consideration are subjected to. While still prone to error, these processes are largely successful in maintaining a high standard of quality.

Common Journal Article Types

As is the case with any newspaper or magazine, many different types of journal articles exist. Also similar to other newspapers and magazines, the bulk of the articles comprise the latest news – original research articles. These articles detail experimental studies and analysis resulting in new knowledge and are arguably the most important articles for both doctoral students and established researchers alike. The process of writing an original research article is detailed in WRITING AN ORIGINAL RESEARCH MANUSCRIPT. Journals also periodically publish review articles, which effectively provide a moving synthesis of a research field, summarising the latest studies and advancements in a single document and acting as an introductory reference for researchers entering the field. The process of writing a review article is detailed in LITERATURE REVIEW MANUSCRIPTS. The publishing process for both original research and review journal articles is virtually the same, encompassing submission, peer review, revision and proofing, and is detailed in the PUBLISHING JOURNAL ARTICLES section of this guide.

Journals also publish variations of these types of articles, the names and format of which may vary from journal to journal. While the elementary scope of this guide makes the discussion of these more niche and less frequently published articles unnecessary, much of its content applies equally to these other article types.

- Shorter original research articles, such as communications, letters or short reports, case studies, methodologies, or methods.
- Review articles with much more room for author opinion, speculation, and hypothesis such as perspectives.

- Articles relating to journal operations, such as letters to the editor, errata (formal corrections to printing mistakes), and corrigenda (formal corrections to published content).

Journal Reputation and Metrics

There are close to 30,000 journals currently in existence from countries all across the world. Many well-known publishers, such as Elsevier, Springer-Verlag, John Wiley and Sons and Taylor and Francis run several thousand journals each. With such a large number of journals being available for submissions, it is perhaps fairly obvious that the status of journal articles as prestigious scientific texts does very much depend on the quality of the journal that they are published in.

The factors that constitute a high-quality journal are somewhat subjective. To many it is the impact factor or other related metrics. These metrics are typically based on the number of citations received by the articles published in a journal compared to its output over a given time period. For example, the impact factor is calculated by dividing the number of citations a journal receives in its reporting year, i.e., the number of times other journal articles refer to content published in the journal and include them in their own respective reference lists over the last year, by the total number of articles that the journal has published in the last two years. For other academics, factors such as the reputation of the journal editor or the publisher are important. Irrespective of how journal quality is determined, the result is the same – a very large number of academics attempting to publish in a comparatively small number of prestigious journals. The time and resources invested in the production of a journal article are considerable and due consideration should be given to publishing a scientific manuscript in a journal and format that will justify that investment. Suggestions for selecting a reputable journal are provided in PUBLISHING JOURNAL ARTICLES — SELECTING A JOURNAL.

Always be on the alert for predatory journals. Theoretically anybody can start a journal, and some use this opportunity to solicit money from inexperienced researchers by offering to publish their research for a fee. Even if no costs are involved, these journals are seldom if ever peer reviewed and have such poor reputations that the work published in them may be discredited. A Google search of the journal name and publisher can help to identify predatory journals. Lists of predatory journals are also available online, such as that compiled by librarian Jeffrey Beall (www.beallslist.net).

Journal Article Access and Licensing

Journal-based scientific communication can be hindered by article access restrictions. Most publishers operate on a subscription-based platform by default. This means that access to articles published in most journals will typically be restricted to customers who are willing to pay for the content directly or in the case of university students and employees, are covered by a university subscription typically managed by its library. These fees help to cover the editorial costs incurred by the publisher while publishing scientific work – as there is typically no requirement for the authors of an article to pay for it to be published. For this reason, the authors will usually relinquish their copyright on the manuscript, signing it over to the publisher during the publishing process.

While websites such as Sci-Hub (www.sci-hub.se) somewhat controversially offer free access to most journal content, it is worth noting that university subscriptions do often offer broad access to content across a range of journals and help to financially support publishers. It is also worth noting that authors are able to share copies of their manuscripts with interested readers and are often only too happy to do so upon request through email or services such as ResearchGate (www.researchgate.com).

Articles which do not require a subscription to view are known as open access articles. Most publishers offer authors the opportunity to publish their article in an open access format if they wish. However, in order to do so, the authors must pay a fee typically ranging from hundreds to a few thousand US dollars. These costs are sometimes covered by the university library if their employees hold a specified authorship position on the manuscript. Open access articles are available free of charge to everybody with internet access and are typically released under a Creative Commons license.

Supplementary Material

Journal articles may be accompanied by additional documents known as (electronic) supplementary or supporting material or information. This content could comprise additional figures, tables, photographs, or micrographs that could not be included in the main manuscript due to journal content restrictions or reduced relevance or importance to the manuscript's story and message. Additional text detailing methodology, equations, results, or discussion and references may also be included. Supplementary material is usually referred to in the manuscript in a similar fashion to figures or tables within the text e.g., S1 (text section appearing in the supplementary material), FIGURE S1 or TABLE S1 (appearing in the supplementary material), or more generically SUPPLEMENTARY MATERIAL once all supplementary material content has already been independently referred to within the manuscript text.

Supplementary or supporting material is submitted alongside the manuscript to the journal by the authors. Document type and formatting is at the discretion of the authors and is rarely modified by the journal. This content is usually available for download on the same webpage as the manuscript and even in subscription journals is free of charge to access.

CONFERENCE PAPERS

Conferences are large gatherings of students, academics, and industry representatives, collectively known as 'delegates', in a single place to exchange information, explore new ideas and socialise. Although some students may be exposed to the conference environment as early as during their bachelor's degree, attendance is typically more relevant to master's and doctoral research degree students, for whom participation in scientific dialogue is considered an important aspect of their programme.

Conferences can be simple or extravagant and may be located close by or require international travel. They are generally themed and can vary in size, although they often have an attendance of hundreds or even thousands of people. Smaller scientific gatherings that are more focussed in their theme and attendance are known as symposia.

Scientific Programme Structure

A conference will typically comprise a subset of themed sessions into which oral presentations will be organised. These are then supplemented with social events and poster sessions. At larger conferences it is not uncommon for several presentations on different topics to be held concurrently in different rooms, with attendees forced to choose between sessions based on their interests and in consultation with the conference programme. Each session is managed by a designated person, known as the 'chair', and typically contains a featured 'keynote' presentation from a prominent invited speaker, which introduces its theme, followed by related oral presentations from other speakers. Especially eminent speakers are also invited to give 'plenary' presentations, which are a small number of extended talks from highly respected members of the scientific community that stand alone as events without any parallel presentations and are usually fully attended.

Conference Presentation Format

Registering to present at a conference is typically completed by submitting an abstract to the conference organising committee via the conference website or by email. Submission timeframes indicated on conference websites are prone to variation with deadline extensions common. When submitting an abstract for approval it is typically necessary to indicate if you prefer to give an oral presentation or a poster presentation and to which conference theme and session your presentation belongs. The rate of acceptance for conference applicants is typically high with most offered either an oral or poster presentation.

An oral presentation typically constitutes speaking in an allotted timeslot with the assistance of a self-prepared PowerPoint presentation and answering questions from the audience at the end. It is important to remain within the allotted time as conference sessions are often carefully scheduled and running over time may result in a rebuke from the session chair. Additional time is allocated at the end for questions, so it is not necessary to dedicate presentation time to this. A poster presentation usually involves taking a self-prepared poster to the conference, mounting it in the designated communal space and discussing it with other conference delegates during session recesses and breaks. Many companies will opt for this option or rent space in communal areas to showcase their products.

Conference Costs

The conference fee is the same regardless of which session or presentation type you register for. Conference fees can range from hundreds to thousands of US dollars and typically depend on the location and venue. For example, conferences that take place in expensive venues such as large conference centres or even cruise ships are generally more expensive than those hosted by universities. Always be guided by colleagues and peers when it comes to conference selection. Conferences are a business like any other and some may attempt to supplement poor scientific content with enticing holiday destinations.

The conference fee typically only covers conference attendance with travel, food, accommodation, and sightseeing costs excluded, although some conferences will provide add on packages that may include some of these extras. Costs incurred by students attending conferences are covered by most universities, although some refuse to reimburse students who elect to give a poster presentation rather than an oral presentation. Conference fees are not charged until an abstract has been submitted, accepted by the organising committee and the delegate has confirmed their attendance.

Opportunities for Conference Delegates

Although students attending a conference are generally expected to present in one way or another, it is also possible to attend a conference as a spectator. Guests may also accompany a delegate, although a fee for the guest's attendance usually applies, and they are generally excluded from the scientific programme with their access restricted to the social aspects of the conference.

Conferences provide students with the opportunity to practice their presentation skills, engage with a wide range of relevant content from international speakers and socialise with their peers, be it students or established academics. They can be a great place to get new ideas, meet potential collaborators, approach academics for supervision, employment, or research visit opportunities, get feedback on ideas or ask questions. For a lot of students, they also provide a very welcome adventure and the chance to travel and explore new places. Students are not technically required to attend any aspect of the conference at all beyond registration, however, conference attendance can be expensive, and the funding body is unlikely to appreciate students that treat conferences as holidays.

Conference Publications

Once a conference submission has been accepted, a PowerPoint presentation and associated talk or a poster should be prepared. A template may or may not exist to assist in this process. A document known as the conference proceedings or book of abstracts will also typically be produced and published following the conference. This is a collection of conference papers covering content presented at the conference, for proceedings, or abbreviated papers or abstracts for a book of abstracts.

Submission of a conference paper with an oral or poster presentation, if allowed, is usually optional. These short papers of typically 10 pages or less, are similar in style to a journal article but are completed on a template supplied by the conference organisers, not normally peer reviewed beyond the initial scrutiny of the abstract during the application to attend the conference and are not copy edited. Proceedings published by reputable conferences are normally indexed by Scopus and other journal indexing platforms but are rarely as widely read or cited as journal articles. That said, some conferences have partnerships with journals and may offer opportunities for the conversion of selected conference papers into journal articles. They may also establish a journal special issue based on the conference which delegates are invited to contribute to.

Use the instructions in Writing an Original Research Manuscript — Abstract to prepare an abstract for a conference submission and Writing an Original Research Manuscript to write a conference paper should one be required for the conference proceedings.

BOOKS AND BOOK CHAPTERS

Hundreds of thousands of academic books are published every year. Most people are familiar with the textbook formats that serve as teaching support materials for primary school through to university students, but many may be less aware of the substantial market in specialised academic references written each year on topics of academic or industrial interest. These books comprise a collection of focussed chapters relating to a central theme and often target highly specific audiences.

Regardless of the theme, the emphasis of a book and its chapters is usually review, a synthesis of information, although authors do sometimes supplement this with their own experimental data. Book chapters generally use a format similar to a literature review but also incorporate fundamentals that may be omitted from other reviews as assumed knowledge. Some book content may be considered too trivial to be published in any other format. Exceptions do exist, especially in sole author books where there is an author vision or greater purpose achieved over the course of the book but in general the novelty and innovation factor in books is low.

Book and Book Chapter Costs

Academically published books, especially on niche topics which do not warrant large print runs are often expensive for interested readers to buy, sometimes costing hundreds of US dollars. Even electronic copies, for which chapters can usually be purchased individually, are expensive and are often not included in university access subscriptions. These factors coupled with the typically established and fundamental nature of book content provides them in a research sense at least with fairly low academic value compared to journal articles. They are generally not as widely read or cited as journal articles, although exceptions exist.

Books and book chapters rarely hold any financial incentive for the contributors. Especially eminent academics working on manuscripts that will form books with substantial commercial value for publishers may be offered some type of compensation for their work but generally contributors can expect a free copy of the published chapter and discounts on the final book or other academic texts from the same publisher at best.

As with journal articles and conferences, book chapters are also sometimes used by predatory publishers to make money. So, if agreeing to write a book chapter, ensure that it is for a reputable publisher and there are no costs associated with the publication. Authors are not normally required to pay any fees when publishing a book or book chapter.

Book and Book Chapter Authorship

Books may comprise chapters written by a sole author, or each chapter may be written by a different, or multiple different authors. Irrespective of who writes the book the commissioning process is usually the same. A book proposal covering details such as the working title, target audience, market potential and proposed content of the book must be produced and submitted to a publisher. This is either completed by a sole author who intends to write the entire book themself or by an academic who intends to find other authors to contribute chapters and will then edit the book. In some cases, publishers may also approach academics without solicitation, inviting them to author or edit a book.

If a publisher commissions a book that will comprise contributions from multiple authors, the academic editor will identify and invite other suitable authors to each contribute a specified chapter to the book. Authors who agree to write this chapter can in turn invite colleagues to contribute to it as co-authors.

Publisher-Contributor Agreements

Once all contributing authors have been identified the publisher will generate formal written agreements, which the authors are required to sign. These outline the requirements for the contribution, including organisation, form, content, style (often in the form of a detailed guide to manuscript preparation) and due date. Details are also provided on the proofing and corrections process, updates and revisions and contributor rights for content use following publication. Contributors must also confirm that the manuscript content is original, has not been previously published or sourced from the public domain, contains no unlawful content or breaches of intellectual property rights, including reuse of copyrighted content where applicable. All content ownership, print and publishing rights for the manuscript are transferred to the publisher under the same agreement.

Writing and Submitting a Book or Book Chapter

Authors generally have a fair degree of flexibility when it comes to the content included in their book or chapter provided it fits the specified book theme. Nonetheless, if in doubt it is always worth running ideas past the editor before beginning to write to ensure that the proposed content is suitable.

Finished manuscripts are usually submitted to the editor by email. The editor will then review the content and make suggestions for any changes that should be made prior to publication. Manuscripts may also potentially be peer-reviewed by external reviewers. Authors are expected to oblige requests for reasonable changes from the editor or publisher and resubmit their revised manuscript within an agreed upon timeframe. The publisher will then produce proofs for the chapter or book for author approval and then publish the book.

Consider the suggestions in LITERATURE REVIEW MANUSCRIPTS when preparing a book or book chapter for submission.

INDEXING AND AUTHOR PROFILES

Scientific publications, such as journal articles, conference papers, books and book chapters are indexed by online platforms that crawl the internet searching for and cataloguing published content. Interested readers can then use appropriate keywords to find relevant scientific documents and track author academic output. Well known examples include Web of Science, Scopus, and Google Scholar, although many others exist.

This cataloguing function is exceptionally important as it enables scientists to quickly and efficiently review topic-specific academic literature from countless sources that would otherwise take a very long time to find individually. Historically such services did not exist. Academics relied on libraries and hardcopies of relevant publications, making document discovery a very slow, tedious, and often uncomprehensive process. Scientific communication was in fact so poor that it was not uncommon for identical discoveries to be made and published separately years apart with each of the authors completely unaware of the other's research.

Citations

Science promotes novelty and innovation. Duplication of existing research is discouraged, yet scientific integrity dictates that all claims must be proven. If every claim in every publication that built upon an expanding research field had to be proven each time it appeared in text, then scientific progress would be very slow and inefficient. References to other publications are subsequently used to substantiate content in a scientific manuscript that has already been proven. Novel findings are supported by practical or theoretical experiment. If something has already been done, there is no need to do it again – readers are simply directed to the existing proof.

The reference from one publication to another is called citation. Every time a scientific publication appears in the reference list of another scientific publication it receives a citation. These citations are recorded by the indexing services, which also scan document reference lists during cataloging. Considerable differences exist in the coverage of the various indexing services. Google Scholar indexes everything including webpages while Web of Science and Scopus are far more selective in their indexing. Subsequently, citations that are recognised by some indexing services may not be recognised by others. Indexing is also a dynamic process and inclusion and exclusion of publications by indexing services can change over time.

Self-citation

Self-citation, or the referencing of one's own scientific publications in other documents, is a common practice and citations obtained in this way are recorded in an author's profile in the same way as any other citation. However, self-citation should not be excessive. References should be selected based on relevance to the manuscript content. If self-citation is appropriate to support manuscript content, then it is completely acceptable and there is nothing wrong with showcasing one's relevant work on the topic, but self-citation should not be adopted preferentially. Some journals have guidelines for self-citation, especially for review articles, but in general it is unregulated. Nonetheless, if more than 10-20% of the references in a manuscript comprise of the author's own work, it may be a good idea to check if more suitable references from other authors exist that could be included.

Author Profiles

Citations are tracked by indexing services and are visible in the profiles of the authors, which comprise a list of their publications and the number of citations associated with each document. The total number of citations for the author is also displayed. These author profiles are platform specific, so academics often have multiple author profiles, one for each indexing service. Author profiles are free and are in some cases even automatically generated upon an author's first publication (e.g., Scopus), while others must be manually created by the author (e.g., Google Scholar and Web of Science).

Author profiles also often include metrics such as the h- and i10-indices. The Hirsch (h) index is defined as the maximum value of h such that the given author has published at least h papers that have each been cited at least h times i.e., an author with an h-index of 6 has published 6 papers that have each been cited at least 6 times. The i10 index is the number of publications that an author has published that have been cited at least 10 times i.e., an author with an i10-index of 4 has published 4 papers that have each been cited at least 10 times. The h- and i10-indices are indexing platform specific, based on the publication inclusions and exclusions of each individual platform. An author's h- and i10-index can subsequently vary by platform. Although citation and hence h- and i10-index potential varies by research area based on publication and citation rate, these metrics still play an important, albeit field specific, role in author reputation, promotion, and award of funding. They subsequently hold considerable value to many scientists.

Open Researcher and Contributor ID

Open Researcher and Contributor IDs (ORCiDs) are codes that uniquely identify authors in scholarly communication. The use of a number as opposed to a name mitigates problems associated with name changes and cultural differences in name order and abbreviation. Unlike other author profiles, ORCiDs do not track citations and associated indices. They simply catalogue author output and match it with an assigned number rather than a name. They are actively promoted by many journals and are often listed with author names on journal articles as a unique author-specific identifier. ORCiDs are free and can be registered online (www.orcid.org).

WRITING AN ORIGINAL RESEARCH MANUSCRIPT

WRITING ORDER

The most efficient writing order for the sections within scientific documents is generally not the order in which they appear. In this guide, the focus will be on writing the core of the manuscript first: the results and discussion in original research journal articles, thesis chapters or conference papers. The reason for this is that this content forms the core value of the entire document with the other sections playing supporting roles in the presentation of this content.

Think of yourself as an author writing the latest chapter in a book that has already been contributed to for years. The 'Results and Discussion' section of your manuscript is your actual product, your ideas and work, and the other sections are the packaging and accessories that help make your product more accessible to your audience.

The 'Introduction' provides background information, scope and context, the problem definition, state of the literature to date, research gaps, research aims and a brief introduction to the materials and methods used to address them. Think of it as your summary of the story that has been told in the book so far, the contributions of other authors, your intended contribution, how it is different, why it is important and how you intend to complete it successfully.

The 'Materials and Methods' section provides instructions on how experimental work was performed. Think of this as a technical supplement to the book. A recipe where you record every detail of what you did and how so that other people can replicate your work should they wish.

The 'Conclusion' section summarises the key findings of the results and discussion, their impact and offers future perspective. Think of this as the ending of your section of the book. Your opportunity to highlight the most important aspects of your contribution and their significance in the greater context of the scientific world and in some cases your ideas for the future.

The 'Abstract' summarises the entire manuscript. Think of it as the blurb of your section of the book. The text which will allow potential readers to figure out what you did, how and why you did it, what the outcome was and why that outcome matters.

Given the reliance of all these other sections on the results and discussion, it is fairly clear that it makes sense to write this section first.

OUTLINE OF RESULTS AND DISCUSSION

Before beginning to formulate any text for the manuscript, it is critical to determine which content to include, in what order it should appear and how best to support it with visual representations of the experimental data. As discussed, concision, precision and logical flow of ideas are the most important aspects of scientific texts, and these can be very rapidly compromised when writing without a structured plan.

Considering the visual representation of data, as graphs, schematics, and tables, at this point is also important as these elements make up the core value of a manuscript, convey the majority of its data and are often the focus of the reader's attention. Manuscript text is most efficiently and accurately built off figures and tables and not vice versa. Hence the planning stage focuses on a flow of ideas and assigning figures and tables to the content.

Consider, as a simple hypothetical example, a study which investigated the effect of aging on the shelf life of untreated and diimide treated (hydrogenated) natural rubber balloons stored for long periods under ambient and elevated temperature conditions. Balloons were manufactured and aged over a period of 24 months at various temperatures. The morphology and chemical composition of the balloon material were then assessed for all samples, and their mechanical properties evaluated.

Begin the outline by listing the variables in the experiment and the measured data from each instrument or output of each analysis.

EXAMPLE - VARIABLES AND MEASURED DATA

Variables	*Measured data*
• Balloon material (untreated and diimide treated). • Storage period (up to 24 months). • Storage temperature (ambient or elevated).	• Composition (elemental analysis and EDS spectra). • Molecular structure (FT-IR spectra). • Morphology (micrographs). • Mechanical properties (tensile strength and modulus). • Thermal properties (mass loss and char residue).

The very first step is to decide on the flow of the manuscript. Think of it as telling a story. What is the data telling you? Look at the variables and measured data from each instrument or output of each analysis: how they are changing with respect to each other; what effects or relationships you can see and what they mean. These are the results. Write these down in a Microsoft Word document as single sentence bullet points.

EXAMPLE - SUMMARY OF RESULTS

- Composition and molecular structure – *samples* (e.g., diimide treated balloons) had a higher *measurement* (e.g., percentage hydrogenation) than the others.
- Morphology – no *samples* (e.g., balloons) stored at *condition* (e.g., ambient temperature) showed any signs of *observation* (e.g., cracking) over the *test duration* (e.g., 24 months). All of the *samples* (e.g., balloons) that were exposed to *condition* (e.g., elevated temperatures) during

aging exhibited *observation* (e.g., micro cracks) with the exception of those treated with *compound* (e.g., diimide).

- Mechanical analysis – no *samples* (e.g., balloons) exhibited any reduction in mechanical properties over the *test duration* (e.g., 24 months) when stored at *condition* (e.g., ambient temperature). The mechanical properties of all *samples* (e.g., balloons) deteriorated when stored at *condition* (e.g., elevated temperatures) except for those treated with *compound* (e.g., diimide).
- Thermal analysis – *samples* (e.g., diimide treated balloons) exhibited greater resistance to thermal degradation than the others.

Group the bullet points into themes. In science these themes are often restricted by instrument type as data must be plotted on an appropriate axis, but instrument results can generally be grouped into broader themes, such as physical properties, mechanical properties, thermal properties, surface properties, morphology etc. which may include results from several instruments. As an illustrative example, results from a CHNS–O elemental analyser and energy dispersive spectroscopy (EDS) could be grouped as both relate to composition. However, to keep things simple the bullet points in the list do not detail instrument groupings.

Decide which bullet points are most important and which are related to each other. What does one tell you about the other? What do your results prove? What is the big picture? The manuscript must have a central message and a story to convey it. Decide on this message based on what the results indicate and write it at the top of the Microsoft Word document.

EXAMPLE - CENTRAL MESSAGE

Samples (e.g., balloons) that are typically *observation* (e.g., thermally and mechanically stable) for at least *period* (e.g., 24 months), are *observation* (e.g., more rapidly degraded) at *condition* (e.g., elevated temperatures) unless they are treated with *compound* (e.g., diimide).

Arrange the bullet points that relate to this central message beneath the new heading, taking care to order them to achieve a logical flow where possible. Sample characterisations that are relevant to multiple sections appearing later in the manuscript — e.g., composition and molecular structure are likely to be relevant for both mechanical and thermal performance — tend to be presented first. However, the other content is generally grouped and arranged based on how one point may relate to or support another, e.g., morphological analysis can often be used to explain mechanical behaviour in materials science. It is also important to consider the figure and table limit at this point and consider grouping data that can be presented in the same figure or table.

EXAMPLE - RESTRUCTURED SUMMARY OF RESULTS

- Sample characterisation: *samples* (e.g., diimide treated balloons) had a higher *measurement* (e.g., percentage hydrogenation) than the others.
- No *samples* (e.g., balloons) stored at *condition* (e.g., ambient temperature) exhibited mechanical property degradation over the *test period* (e.g., 24 months). Morphological analysis showed *observation* (e.g., no cracking) in these samples. Conversely, all *samples* (e.g., balloons) stored at *condition* (e.g., elevated temperatures) exhibited *observation* (e.g., mechanical

property degradation) except for those treated with *compound* (e.g., diimide). Morphological analysis also showed *observation* (e.g., cracking) in all *samples* (e.g., balloons) stored at *condition* (e.g., elevated temperature) except those treated with *compound* (e.g., diimide).

- Thermal analysis indicated that *samples* (e.g., diimide treated balloons) exhibited greater resistance to thermal degradation than the others. We attribute this to the treatment of the material with *compound* (e.g., diimide). This is supported by theory, literature, or further experiments.

Remaining bullet points which do not fit the 'central message' and story can be combined with additional experiments to create a new manuscript or can be included as supplementary material or an appendix accompanying the manuscript. Some bullet points in the results list may require supporting measurements, which may be included in a characterisation section if applicable or in the supplementary material or appendix depending on the manuscript type.

Now, look through the results list and decide the appropriate medium of expression for each bullet point. Small quantities of data can be expressed in the manuscript text using brackets and a value with units. Larger quantities of data can be expressed using a table or figure. A single table or figure may sometimes be able to communicate multiple bullet points if common parameters and units are shared or multiple axes are used.

If the data associated with the bullet point can be summarised using key parameters e.g., maxima, minima, averages, then aim to assign a table or scatter plot to this data. A table can be used to present many variables while a 2D scatter plot is restricted to two variables if using a single set of axes. If the data associated with the bullet point comprises functions that should remain intact e.g., spectra, then aim to assign a graph to this data.

EXAMPLE - DATA PRESENTATION PLAN

- Composition and molecular structure: depending on the number of elements or structures analysed and data format either spectra or a table. A scatter plot could also be used if there was a trend that should be graphically illustrated.
- Mechanical analysis: a graph, a table or both. A graph may be important to provide detailed insight into variable dependent behaviour, e.g., a stress-strain curve, in some samples. Otherwise, it may be appropriate to simply summarise key properties using a scatter plot or table.
- Morphology: micrographs (images).
- Thermal analysis: depending on the relevance of the *measurements* (e.g., temperature ranges analysed) to the line of inquiry (e.g., balloon use and storage temperatures), this could be included as a graph (measurements and line of inquiry are similar and the data is hence important), or key values simply commented on in brackets within the manuscript text to support discussion in other sections with the graph moved to the supplementary material.

Figure and table number restrictions vary but aim to assign 6-8 figures to the bullet points and up to two tables if writing a journal article. Conference papers typically have fewer figures and tables than journal articles. Thesis and book chapters can contain as many figures and tables as is necessary to communicate the chapter content efficiently and effectively but should not look overcrowded. Combine figures and tables where appropriate.

Retain this planned outline of the results and discussion and generate the figures and tables now based on the ideas presented in the next section of this guide.

FIGURE AND TABLE DESIGN

Figures and tables are arguably the most important aspect of a scientific manuscript. They contain virtually all the information collected in the study and comprise the core value of the manuscript, delivering a large amount of data in a comparatively small amount of space. They are in some cases the only part of the manuscript that a reader will look at other than the abstract or conclusion and it is subsequently critical that they are well planned and assembled. They should not contain irrelevant content, should be presented in the most appropriate format — i.e., some types of data are better expressed using one type of figure or table as opposed to another — and should have a very clear and easily interpreted message. A reader should be able to interpret the central message of the manuscript and its key findings from a glance at its figures.

If writing and publishing your own scientific document e.g., a thesis, then there is a lot of flexibility when it comes to document formatting. However, in general, most journals structure content in columns, with more than one column typically appearing on a page. Schematics and graphs subsequently often appear much smaller when published in a journal than they would in another type of manuscript. For this reason, it is important to pay attention to the size of text and features appearing in figures and ensure that they will still be intelligible if considerably reduced in size.

In some instances, permission from the publisher is also required to use colour. Even if a manuscript will be published in colour, many readers may choose to print it in black and white. Subsequently, all figures should ideally be able to be interpreted based only on their geometry rather than relying on colour differences to distinguish data sets.

Micrographs

Scanning electron microscopy (SEM) is a popular tool for morphological characterisation across many scientific disciplines and yields greyscale micrographs (images). These micrographs often provide an engaging visual representation of scientific work and are commonly included in manuscripts to support quantitative data. Several imaging modes can be used. The following suggestions apply to secondary electron micrographs, the most widely used SEM micrograph type. That said, many of the recommendations on micrograph presentation are also relevant to other SEM imaging modes e.g., backscattered electron micrographs, micrographs produced using transmission electron microscopes (TEM), optical microscopes or even photos.

While lower resolution micrographs (e.g., 300 dpi) are often favoured on microscope computers due to their smaller storage requirements, it is always a good idea to ensure that the micrographs that will be included in a manuscript are generated at a higher resolution (e.g., 600-1200 dpi) to allow for processing and publication image resolution requirements.

SEM is not a blanket characterisation method so micrographs must either be taken across multiple locations of a sample and a representative micrograph selected or focus on a specific feature being discussed in the manuscript. The most important aspects of a micrograph are that it is in focus and details the region of interest at the correct scale. Focus is not easily modifiable during post processing and while some degree of scale correction is possible, e.g., increasing the size of various regions by cropping and stretching, only minor adjustments are possible without compromising image quality.

Magnification and Scale

The magnification and scale of a micrograph are not the same thing. A micrograph can be digitally stretched to twice its original size, yielding a 2× magnification, while its scale remains unchanged, with every feature including the scale bar simply twice as large as it was previously (FIGURE 1). Scale is subsequently the preferred reference medium so attention should be afforded to the scale bar rather than the magnification. If comparing micrographs, ideally, they should all be the same scale. It is also critical that the scale is such that any features of interest are clearly visible. All micrographs should include an indication of scale; however, default microscope information bars typically include other details, such as instrument name or branding, accelerating voltage, working distance etc. Manually generated scale bars are subsequently preferred and can be generated using a range of software. Software selection is a matter of author preference; however, this guide uses the Fiji distribution of the freeware ImageJ (www.imagej.net) and Adobe Photoshop (www.adobe.com) or a

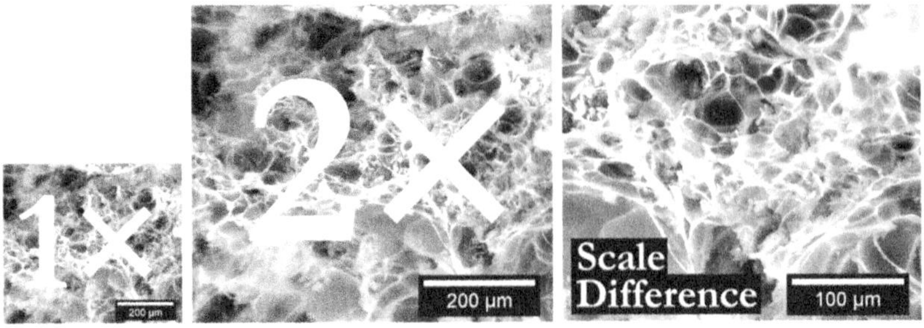

FIGURE 1. *Magnification and scale differences in SEM micrographs. Moving left to right, the first two micrographs have different magnifications while the third micrograph has a different scale.*

lower-budget alternative, such as Paint.NET (www.getpaint.net). ImageJ is excellent for easily cropping images while retaining scale and for generating scale bars. ImageJ also provides options for modifying brightness and contrast. However, the aforementioned specialised photo editing software are usually easier to work with and yield better results.

Brightness and Contrast

Since micrographs are greyscale, it does not matter if they are first cropped and their microscope information bars removed or their brightness and contrast first modified. However, if using ImageJ for cropping and scale bar generation and other photo editing software to adjust the brightness and contrast, then it makes sense to begin with the latter. This is because, once the microscope information bar has been removed, the scaling of the micrograph is no longer apparent. It is subsequently more efficient to first modify brightness and contrast in the photo editing software, save the images and then open them with ImageJ. The scale of the micrographs can then be set, the images cropped and artifacts or the microscope information bars removed, and the final scale bars added in ImageJ in one step without losing track of the scale. The scale bar should not be added until the microscope information bar has been removed since it will tend to overlap with this bar if generated too early and make cropping challenging.

Once suitable micrographs have been obtained, modify the brightness and contrast in your preferred photo editing software to generate clear and engaging visual impact for the manuscript. Start by increasing the contrast until the micrograph is almost overexposed and the majority of shades are represented as blacks or whites (FIGURE 2). Then lift the brightness to reveal regions lost to the increased contrast, dialling the contrast back as necessary if the micrograph becomes overexposed. The darkest point on the final micrograph should be completely black and the brightest part completely white. It should have a defined black and white appearance rather than an overall grey tone; however, if sizable black regions are present be sure to increase the brightness to check these regions for lost features. This is especially important in micrographs with considerable differences in depth of field, where some regions tend to be very dark and others very bright. If possible, avoid such differences in depth of field and select more uniform sample regions as capture sites. When you are satisfied with the micrograph brightness and contrast, save it as a new image file.

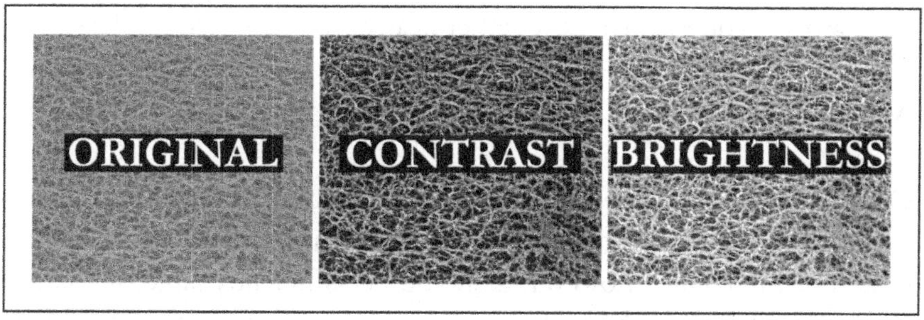

FIGURE 2. *Modification of the contrast and brightness of a SEM micrograph using Paint.NET. Contrast is increased to bring out the blacks and whites in the micrograph and the brightness is then lifted to reveal lost regions.*

Cropping Micrographs

The micrograph can now be opened in ImageJ and the microscope information bar removed. Cropping can also be used to remove artifacts from the micrograph or to focus on a specific region, although care should be taken that the micrograph resolution is sufficient to permit cropping without considerably compromising image quality.

Before cropping it is important to set the image scale so that a new scale bar can later be generated, which accurately reflects the scale of the micrograph without reference to the original microscope information bar. Begin by using the straight-line tool to draw a line along the length of the scale bar on the microscope information bar. Then select 'Analyse' and 'Set Scale'. Enter the scale bar length in the 'Known distance' field and its unit in the 'Unit of length' field before clicking 'OK'. The scale of the micrograph is now set, and it can be cropped at will including removing the microscope information bar without losing the scale.

Crop by highlighting the region that should be retained using the rectangular selection tool. This could be all contents of the micrograph other than the default microscope information bar, areas of the micrograph not affected by artifacts or a specific region of interest. Then select 'Image' and 'Crop'. If the selected region is too small, then there will be a considerable reduction in image quality and new micrographs unaffected by artifacts and of a more appropriate scale should instead be generated. If multiple micrographs will appear together as part of the same figure, then care should be taken to ensure that they are cropped to approximately the same aspect ratio and will fit together neatly.

Adding a Scale Bar

Once the micrograph has been cropped to the desired dimensions, a single scale bar should be added (FIGURE 3). Scale bar styling is a matter of author preference; however, it should be sufficiently large that it will still be intelligible if the micrographs are resized during the publishing process (for journal articles) and should be starkly contrasted against the background colouring of the micrograph. To be on the safe side, a larger scale bar with a background colour of its own e.g., a white scale bar with a black backing overlaying the micrograph, can be used. If no background colour is used for the scale bar, then there should be considerable contrast between the micrograph contents and the scale bar colour e.g., if the region of the micrograph where the scale bar will appear is light in colour then use a dark scale bar and vice versa. Use of clear labelling on the scale bar to indicate its length improves clarity, however, if desired numbers can be omitted from the scale bar with its length instead noted in the figure caption. Some authors also simply generate a white space within the micrograph, which correlates with a specific length rather than generating a bar.

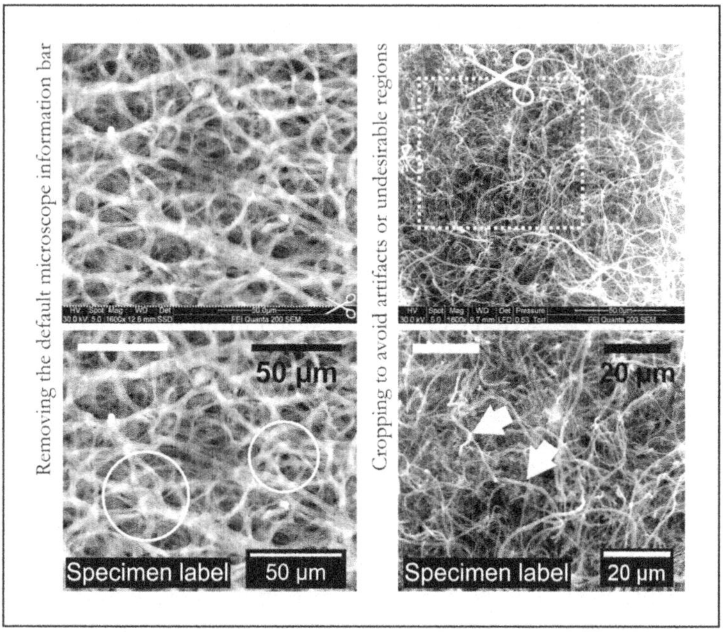

FIGURE 3. *Cropping micrographs to remove the default microscope information bar or avoid artifacts or undesirable regions (top row) and creation of a scale bar and labelling (bottom row). As illustrated, scale bars can be black or white, with or without numbers or overlay. Notable features can be indicated with circles or arrows.*

To generate a scale bar in ImageJ, ensure that micrograph scale and unit have already been defined, as described earlier, and then select 'Analyse', 'Tools' and then 'Scale Bar'. There are then options available for the aforementioned scale bar types. When setting the width of the scale bar try to select nice round numbers, such as 5, 10, 20, 50, rather than random numbers, such as 13, 22, 45. The ideal sizing of the scale bar relative to the micrograph really depends on the scale of the features of interest but in general try to keep the scale bar width to roughly one-fifth to one-third of the micrograph width. Smaller scale bars may be difficult to interpret if resized during publishing and larger scale bars may be difficult to use unless the features of interest occupy a large portion of the micrograph area, and the scale bar is intended to representatively quantify their size. Scale bars will typically be located in one of the corners of the micrograph, but their location is best selected based on the premise of not visually obstructing important features within the micrograph.

Label directly onto micrographs to provide clear and obvious insight into the most important information and features for readers. Use black or white circles to highlight certain regions, large arrows to indicate specific features and text on the micrograph or in the figure caption to describe micrograph content.

A second micrograph with a different scale can be overlaid in a corner of a micrograph where necessary to form an inset, illustrating phenomena within the same region occurring at different scales. These should, however, be sized such that they are still discernible if reduced in size by a publisher and should have their own scale bar detailing the new scale.

Small artifacts (faults in the micrograph such as charging on edges etc.) and differences in micrograph dimensions or aspect ratio are not especially problematic provided there is sufficient space to crop the micrograph accordingly. Optimal regions can be isolated from a micrograph, although care should be taken to retain scale and dimensional consistency both within a micrograph and across micrographs appearing together in a manuscript. With the exception of cropping, micrographs should not be digitally altered in any way as this may constitute data manipulation. Although micrographs can be artificially coloured for artistic purposes they should not be coloured for scientific publication.

Graphs

Graphs are visual representations of data and can make the meaning associated with numbers more intuitive to readers. They are typically produced digitally with the assistance of software. There are many commercial providers of graphing software and selection is a matter of personal preference. Attractive and informative graphs can theoretically be produced using any software package if the user is competent. Microsoft Excel (www.microsoft.com) is perhaps one of the most popular pieces of software for graphing among students. However, software such as that produced by OriginLab (www.originlab.com) tends to be more widely used by established academics. The graphs presented in this guide are generated using OriginPro. Most universities will have provisions to offer complimentary or discounted software to students for personal use.

If such provisions are not available, then relevant software is usually installed on university computers.

Graph Selection

Every graph has its own advantages, disadvantages, and most suitable usage scenarios. In most cases, however, graph selection is fairly simple and unambiguous. The purpose of a graph is not to look impressive or complicated. It is to represent data and relationships clearly and accurately. The author's intended message and the purpose of the graph should be immediately obvious at a glance. Strength and accessibility of the 'main message' is the priority. The main message in each figure should correlate with a key point from the 'Restructured Summary of Results' and the figures, when presented together, should convey the 'central message' of the manuscript. Data that should be included in the manuscript but does not contribute to the 'central message' should be moved to the supplementary material (for journal articles) or an appendix (for theses). Alternatively, this data can form the basis for another manuscript.

Scatter plots and line graphs are often excellent choices for figures (FIGURE 4). Scatter plots are ideal for distinct data points that do not form part of an ongoing complete data series. Points are only joined by connecting lines if they represent intermittent measurements in an ongoing series i.e., a representative but incomplete data set where additional points exist between the measurement intervals that are approximated using the connecting lines. These graphs present information in a clear and efficient manner. Trends are immediately visible, and values easily ascertained. In contrast, column graphs and bar charts are highly inefficient, consuming considerable space that could be occupied by other data series with effectively irrelevant colouring or shading, which also consumes a lot of ink if printed. Column graphs are easily replaced with more efficient scatter plots. To generate a scatter plot in OriginPro, select the relevant data, right click it and choose 'Scatter' from the 'Symbol' sub menu of the 'Plot' menu. If connecting lines should be present, select 'Line + Symbol' instead.

Visually appealing 3D graphs can be exceptionally difficult to read and even harder for readers to extract tangible data from (FIGURE 5). They are

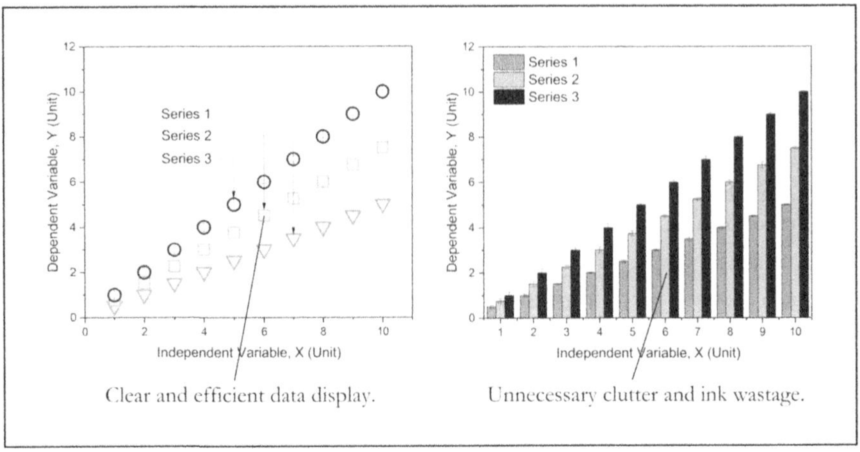

FIGURE 4. *A single data set presented using a scatter plot and column graph. Both deliver a message and are interchangeable, but the scatter plot is much less cluttered. Colours appearing in this figure may not be interpretable on some devices and in some print formats.*

generally only useful to provide visual approximations of trends between multiple parameters.

Grouping Graph Content

Use graph space efficiently, grouping as many relevant data series into a single graph as appropriate (FIGURE 6). A graph with too little content is a wasted opportunity to include results in what is generally highly restricted manuscript space i.e., most journals have a figure limit. On the other hand, strongly avoid including data that is not important or relevant to the message that the figure is attempting to convey as this will potentially confuse and misdirect readers. Even if multiple data sets are included in a figure, a main message or theme should be present, supported by the showcased results.

If the number of figures should be minimised to meet journal requirements, it is acceptable to group figures together to form a multi-part figure, i.e., multiple related graphs labelled with letters identifying their place in the group and appearing together in the manuscript as a single figure. This can be a good alternative to overcrowded graphs, graphs with too many axes or graphs displaying data that is similar in theme but

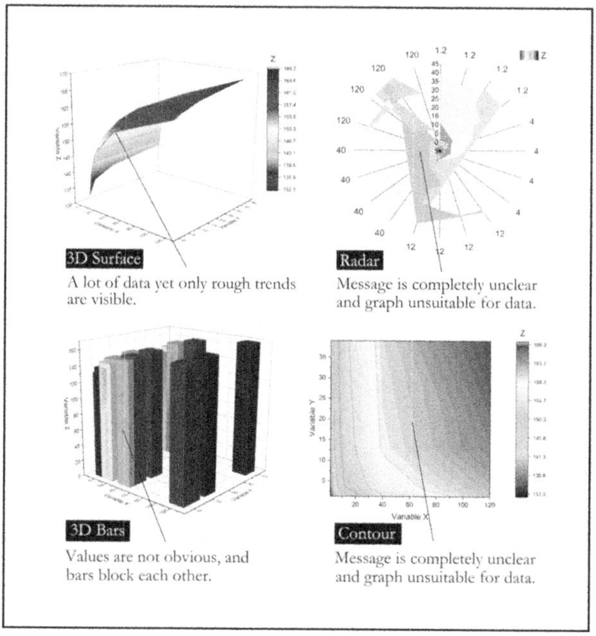

FIGURE 5. *A series of complex graphs (3D surface, radar, 3D bars, contour) all produced using the same data set. All look great but do not simply and effectively deliver a message. Colours appearing in this figure may not be interpretable on some devices and in some print formats.*

should not appear on the same set of axes due to differences in units. Multi-part figures that use a common axis, i.e., either the horizontal or vertical axis has the same parameter, scale, and unit in more than one part of the figure, can be arranged horizontally or vertically adjacent to each other and the axis labels removed from the parts of the figure occurring after the appearance of the first fully labelled axis. This allows the spacing between them to be a lot narrower and looks neater. All axes should, however, still contain increment markings. Size and arrange grouped figures to have a straight edge on all sides so that when combined they form a neat square or rectangle.

Graph Layout Basics

Graphs comprise axes on which variables can be plotted. A variable is simply an attribute that changes, potentially with respect to something else.

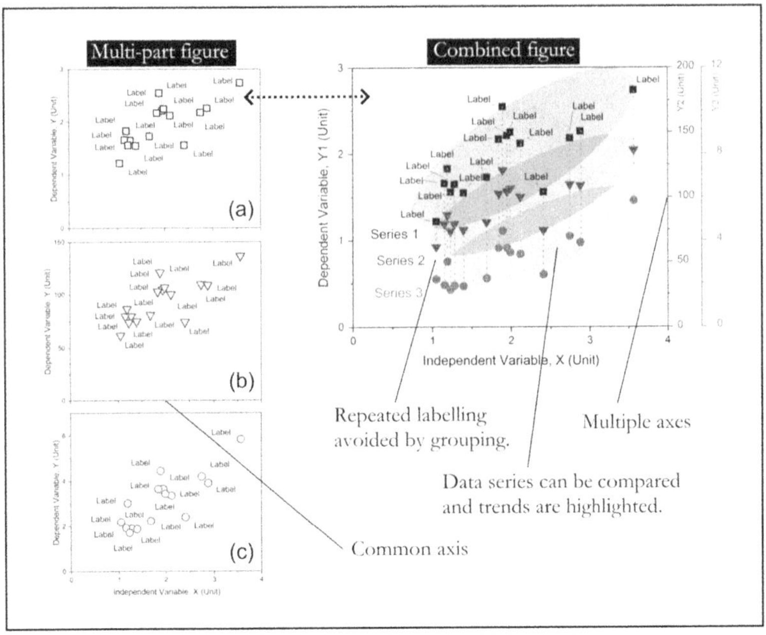

FIGURE 6. *A multi-part and combined figure displaying the same data sets. Multi-part figures are less cluttered, but data is often easier to compare in a combined figure. Colours appearing in this figure may not be interpretable on some devices and in some print formats.*

An independent variable is an attribute that causes a change e.g., percentage hydrogenation of balloon material and is usually plotted on the horizontal X axis (FIGURE 7). The dependent variable is an attribute that changes as a result of the independent variable e.g., the tensile strength of the balloon material and is usually plotted on the vertical Y axis.

While sample labels may seem the simplest and most intuitive basis for X axis labelling this should be avoided where possible. Plotting measured properties against the labels of the respective samples robs the graph of potential accessibility and impact. The sample label itself is not an intuitive variable. When sample names are used for axis increments instead of a single defined variable the graph's main message is limited to the measured properties in the context of sample variation (which could potentially be based on many variables). Even if the sample labels do reflect changes in only a single variable, a numbered axis is more intuitive for readers than trying to figure out the meaning behind sample labeling,

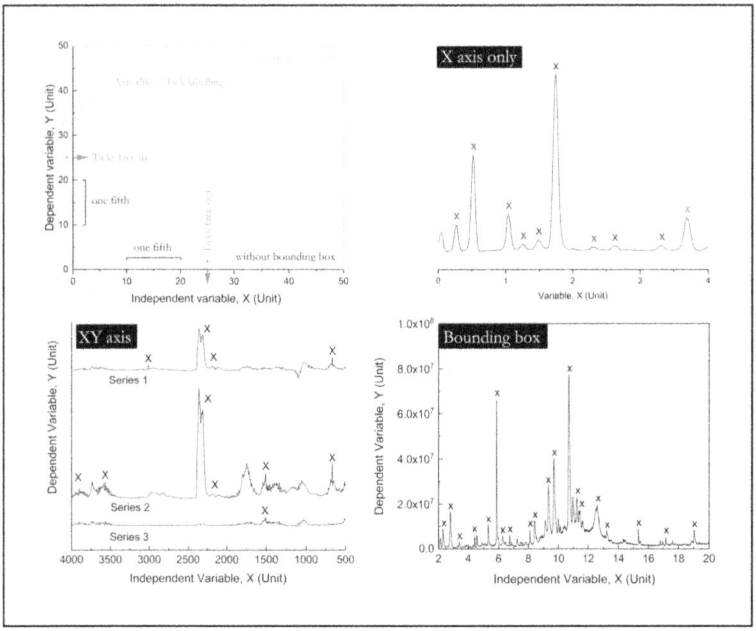

FIGURE 7. *Graph layout options and examples of plotted spectra. Appropriate axis, tick and labelling options must be selected. In some cases, only the horizontal X axis is important and the vertical Y axis is omitted. Graphs can optionally be closed with a bounding box.*

how it varies and how this is in turn affects the measured properties. Authors and not readers are responsible for performing a comprehensive and complete analysis and providing the results in an obvious and easily interpreted fashion. Ensure that you establish the variable behind the sample label and plot the measured properties with respect to this variable rather than against the sample label e.g., the X axis indicates the percentage of a component in a blend rather than a series of labels detailing the blend names.

Label graph axes with a parameter description written in full, followed by its associated symbol (if applicable) and unit in brackets. A graph displaying data for which only a single variable is important, as is the case for some spectra, can comprise a data series located above a single horizontal axis. Graphs may be closed with a bounding box or left open based on author preference. Default font sizing can be retained in OriginPro (22 pt. for axis titles, 18 pt. for tick labels). Tick labels can be formatted to face in or out. This is a matter of author preference; however, inward facing ticks can sometimes obscure data.

Ensure readability of graph axis labels and numbering through adequate font sizing. Numbering should not be considerably smaller than axis labels – one font size increment is usually enough. The division of an axis into labelled increments of one-fifth is generally sufficient to provide ready accessibility to data values without crowding the axis with labelling. Increments should ideally be round number intervals e.g., 10, 20, 50, 100. Use scientific numbering e.g., 1×10^6 to express very large or very small numbers. Specific graphs may require finer numbering increments.

All axis options can be configured in OriginPro by double clicking on the relevant axis and choosing the applicable options from the 'Scale', 'Tick Labels', 'Title' and 'Line and Ticks' tabs.

Using Multiple Axes

If many parameters are to be plotted on a 2D graph, multiple horizontal and vertical axes can be used, one on each of the four sides of the square or rectangular space (FIGURE 8). Additional axes can also be added with an offset. Use of multiple axes is only justifiable when that allows multiple parameters with different units to be presented together with the intention of their comparison and the expression of a clear message or messages pertaining to that comparison. If this is the case, ensure that axes are scaled appropriately to sufficiently separate the data series in question.
Corresponding axes and data sets should be carefully formatted or labelled so that it is always clear on which axis a data series is plotted. This can be achieved by labelling each data series with the axis title or abbreviation and using the same colour for both. Use of an abbreviation and wider spacing of ticks can allow the axis label to be recessed and the multiple axes to be positioned closer to one another.

Additional axes can be added in OriginPro by adding an appropriate layer. These can be 'Bottom X Left Y', 'Top X', 'Right Y' or 'Top X Right Y' axis layers. Once the desired axis has been inserted, data series can be added to it by right clicking the graph layer in the object manager and selecting 'Plot Setup'. Then simply ensure the new layer is selected and assign the relevant data series X, Y and associated error attributes using the check boxes before clicking 'Add' followed by 'OK'. If data plotted on

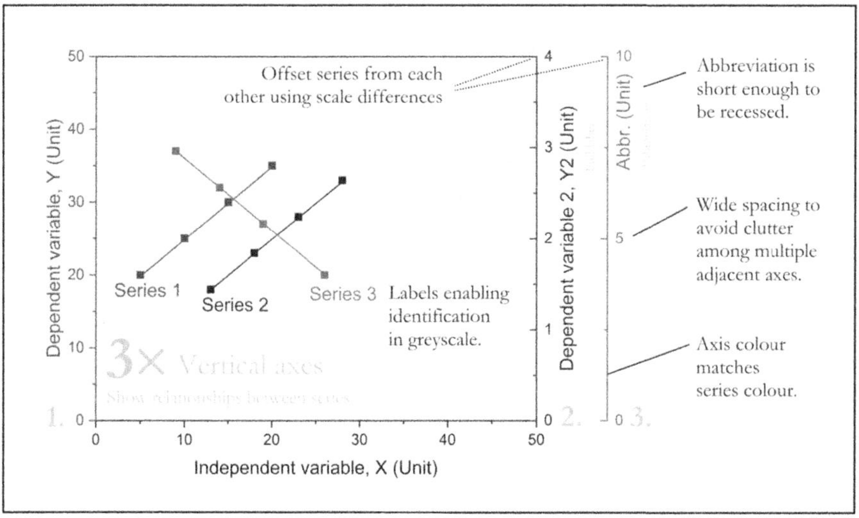

FIGURE 8. *Layout options for a graph with multiple axes intended for the comparison of data with different units and scales to provide a clear message that would not be apparent using another graph type. Use different axis ranges and tick intervals to arrange data sets neatly with respect to each other. Colours appearing in this figure may not be interpretable on some devices and in some print formats.*

the new layer is not scaled correctly then be sure to check the scales of the axes in the different layers by double clicking an axis and cycling between the layers using the drop-down box in the lower left-hand corner of the settings window.

Line and Point Formatting

Use line and point formatting to clearly distinguish data series from each other. This can be achieved using hollow and solid markers, markers of different shape, solid lines and dotted or dashed lines of various patterns (FIGURE 9). Different coloured lines and markers can also be used to distinguish data series but should not be relied on as the only characteristic factor due to the potential for the manuscript to be printed in black and white.

Line and point formatting can also help to make the message of the figure clearer. Relationships between series can be indicated through common formatting, e.g., all hollow markers, all dashed lines for related samples.

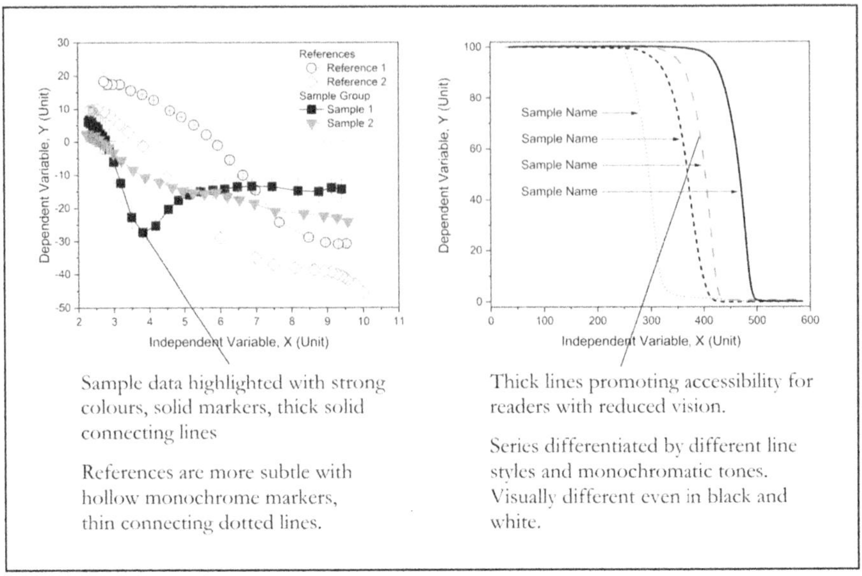

FIGURE 9. *Distinguishing and grouping data using line and point formatting, such as different or similar colours, shades, shapes, hollow and solid markers and solid, dotted or dashed lines. Colours appearing in this figure may not be interpretable on some devices and in some print formats.*

References can be black or grey and test data coloured to make their identification clearer. Desirable results can be marked in green while less desirable results can be marked in orange and red.

Maintain consistent formatting across all figures in the manuscript to enable readers to familiarise themselves with your conventions and more efficiently interpret data and results. Always use identical wording in labelling when referring to the same data or samples. If specific data series or samples are formatted in one figure using a certain marker or line shape, pattern or colour, use identical formatting for the same data series or samples in the other figures appearing in the manuscript.

Thicker lines and larger marker sizing make graphs more accessible to readers with reduced vision. Their use can, however, exacerbate overcrowding problems and affect accurate interpretation of data due to the much coarser visual representation that they provide. Hollow markers can be preferentially used over solid markers at larger sizes as they often visually contribute to overcrowding less than solid markers would and make error bars easier to see. Data series with a large number of points or overlapping points can also look less cluttered and more easily interpretable when hollow markers are used.

All line and point formatting can be configured in OriginPro by double clicking the relevant data series and choosing the desired options from the 'Symbol' and 'Line' tabs. By default, the 'Dependent' edit mode is enabled on the 'Group' tab. While this is enabled, the data series are formatted as a group. To format each data series independently make sure the 'Independent' edit mode is selected on the 'Group' tab.

Representing Error

Error is an inherent part of science. Every data point should be based on multiple replicates of the same experiment with the average value plotted and a measure of the error, such as standard deviation or standard error, represented using error bars. An exception exists for spectra or other measured data forming a continuous series that is highly sensitive to incremental variations in the independent variable e.g., FT-IR spectra, stress-strain curves etc. Averaging data with such potentially large differences in the data values across replicate samples often results in a completely inaccurate average. Instead, plot all replicate data and choose a single 'representative' data series that best depicts the behaviour of the other replicates for use in the final graph. That said, appropriate properties or attributes that are derived from such data can be expressed as an average and measure of error e.g., while a series of stress-strain curves for replicate samples should not be averaged since their Y values often vary too much with respect to their X values for an average to be an accurate representation of the sample behaviour, the tensile strength (highest Y value) of each replicate can be determined and averaged across the set. Acceptable error margins vary based on the context of the experiment e.g., even very small variations in experimental data may be unacceptable in instrument-based measurements of high purity samples while considerable error may be expected in biological products due to inherent natural variations.

If the error associated with a measurement is too high to be considered acceptable, do not simply select some results and omit others to reduce it. This can be considered data manipulation. Complete further measurements instead until the error is reduced to an acceptable level across a larger sample size. An exception exists for data that can be clearly identified as outliers. Generally, data falling more than 3 standard deviations outside the mean can be omitted as outliers. This may be reduced to 2 standard deviations from the mean for smaller data sets.

Error bars are frequently fitted vertically for the dependent variable and can show the upper limit of the error, the lower limit or both. In general, it is ideal to show both since this represents the full scope of the error. Error bars can also be fitted horizontally if two variables that both contain error are plotted with respect to each other.

Error bars are automatically fitted in OriginPro, provided the workbook contains an X (xEr±) or Y (yEr±) error column containing values. To designate a column as an error column, select it, right click it and choose 'X Error' or 'Y Error' from the 'Set As' menu.

Labelling and Legends

Label directly onto the figure using large clear font and full wording where possible. This enables interpretation of all information in the figure as quickly and efficiently as possible with maximum reader engagement. Readers should ideally not need to search the manuscript to find previously defined abbreviations or acronyms in the text. If the use of abbreviations or acronyms is necessary in the figure due to space restrictions, then they should be defined in the caption accompanying the figure. Arrows can be used to match text with data series. Avoid diagonal arrows. Instead use an arrow comprising a horizontal and vertical translation forming a right angle. All arrow heads should point in the same direction e.g., if one arrow points left and down, then they should all point left and down. Arrows and labels can be added using the 'Arrow' and 'Text' tools in OriginPro.

When labels have words in common e.g., red balloon, yellow balloon, green balloon, then use a list with a single heading in bold, e.g., balloons, and the distinguishing words e.g., red, yellow, green, listed under it to avoid repeating words. Arrows can then run from each distinguishing word to its corresponding data series. Be sure to adjust the axis scale as necessary to optimise data series spacing and make labelling easier and clearer. A smaller axis value range can help to separate data series. A larger axis value range can help to generate blank space at the figure's extremities, making space for labels and allowing larger label text sizes. Axis scale can be configured in the 'Scale' tab of the relevant axis in OriginPro.

When data series all appear very close together and have a similar form, a legend provides a better alternative than attempting to label the series directly. Legends are automatically generated in OriginPro but if deleted or modified can be reconstructed or reset using the 'Update Legend' or 'Reconstruct Legend' options from the 'Legend' submenu of the 'Graph' menu. The text in legends is automatically sourced from the workbook but can be manually modified (both content and font) by double clicking it. The legend fill and border can also be formatted as desired using the 'Fill Color' and 'Border/Line Color' tools. To remove the legend border, use a border line thickness of 0

Improving Data Layout

Sometimes it is necessary to offset the first tick mark, typically associated with zero, of an axis from the perpendicular axis (i.e., offset the first tick on the horizontal axis from the vertical axis or vice versa). This is often the case if there are data points associated with the first tick mark, which if not offset will interfere with the perpendicular axis or be difficult to label. Such points can be offset by entering a lower bound for the axis range that is less than zero but does not exceed the increment value for tick spacing. If it does exceed this value, then rather than the first tick mark being offset from the perpendicular axis, a new tick in the negative region will be generated. In OriginPro axis scale can be configured on the 'Scale' tab of the respective axis.

Axis breaks can also be useful to modify graph scaling for improved data visibility. If data spans a very wide range but exhibits regions where there is very little change in the data values, these sections can be omitted with an axis break (FIGURE 10). This is only acceptable if the regions on either side of the break are representative of the omitted content or the omitted region is completely irrelevant. In adding an axis break the axis space associated with the range is reduced which in turn makes any variations in the other data more obvious due to their greater scale. Axis breaks can be configured in OriginPro on the 'Breaks' tab of the relevant axis.
Graphs can also sometimes include diverse data sets that must be com-

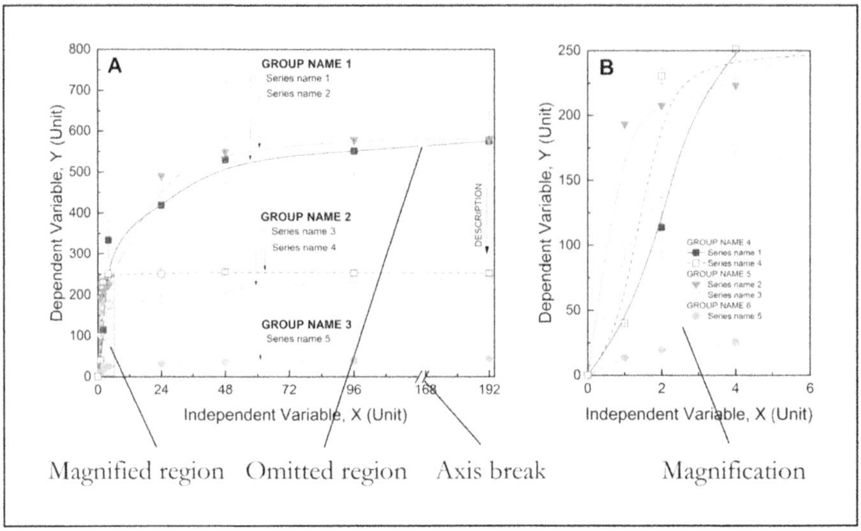

FIGURE 10. *A graph incorporating an axis break and magnified region. These can be used to make small, yet important regions lost to scale differences visible without compromising the scaling that is most appropriate for the majority of the data and the main message of the graph. Colours appearing in this figure may not be interpretable on some devices and in some print formats.*

pared yet vary considerably in magnitude over an extended scale. It is possible that in order to visualise the main trend of a data set it must be scaled in such a way that important features become too small to be noticed. In such cases an inset can be produced to magnify important regions of a graph while still maintaining the most relevant scale for the rest of the data set. To create an inset in OriginPro select the 'Scale In' tool, hold the 'Ctrl' key and drag the mouse over the region that should be magnified. A new graph will be created of this magnified region, which can then be pasted onto the original graph as an inset or adjacent it to form a multi-part figure. The magnified region will also be automatically indicated on the original graph with shading.

Highlighting the Message

Be creative and innovative when it comes to getting the message in graphed data across. A good graph is more than just numbers. Use a hybrid approach during the design process – create a graph but try to incorporate schematic style elements that make the real meaning behind the numbers clearer and more intuitive for readers.

- Regions of interest in data series can be shaded and labelled.
- Trends can be indicated with labelled arrows and explained with symbols or a short one-to-three-word simplified explanation.
- The statistical significance of trends can be established using statistical analysis and fitting. Provide fitted curves and label them with statistically relevant information e.g., R^2 or p-values.
- Small images or schematics can be placed next to data points to help illustrate their meaning.

Straight 'construction' lines can be drawn onto the figure during the generation of new elements to support this process, providing a reference for spacing or alignment of one element with respect to another, and deleted again when the figure is finished.

FIGURE 11 combines picture and graph elements (energy dispersive spectroscopy (EDS) spectra and colour-coded element maps) to illustrate the relationship between the two and add additional depth to the message being conveyed. Relevant peaks in the spectra are linked with the element maps using solid lines and a coloured bubble matching the colour that the element appears in the map. The presence of certain peaks in some spectra is noted and their absence in others highlighted. Additional labeling could also be added to convey important information or trends directly to the reader e.g., high or low element X, Y or Z content etc.

Exporting Graphs

Once graphs are finished, they can be added to the manuscript draft. This can be achieved by saving the graph as an image file and inserting this image into Microsoft Word or by copying it directly across as an editable file. To save a graph as an image in OriginPro select 'Export Graphs' from the 'Export' submenu of the 'File' menu. Then select the appropriate image type, file name and path, resolution and colour settings followed

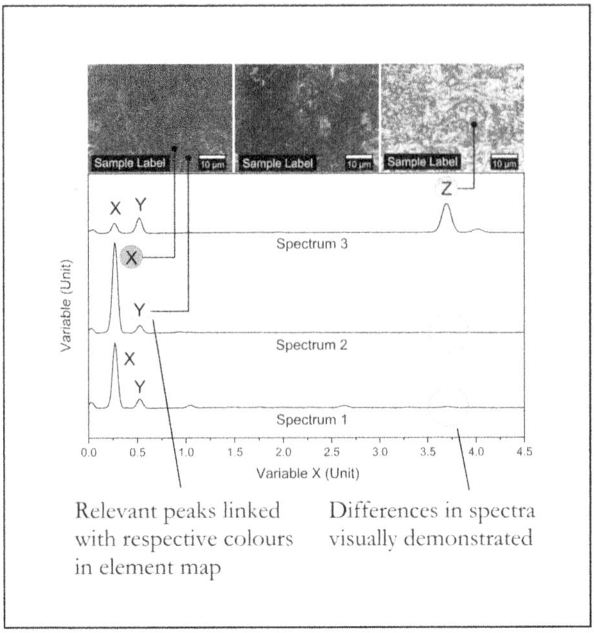

FIGURE 11. *Picture-graph hybrids explicitly highlighting important points and trends make the message that should be relayed to readers more obvious and more intuitive. Use shading, colours, boxes, arrows and labelling to directly communicate with readers. The faster and more easily they get the message the better. Colours appearing in this figure may not be interpretable on some devices and in some print formats.*

by 'OK'. The image itself cannot be further edited in OriginPro once it is exported so it is important to retain the source file.

Alternatively, editable graphs can be pasted into Microsoft Word by opening the graph in OriginPro and selecting 'Copy Page' from the 'Edit' menu. Then paste the content into Word and crop as necessary by right clicking the editable graph and selecting 'Crop'. To edit the graph again just double click it and it will open in OriginPro. The editable graph is embedded in the Microsoft Word document and can be edited by anybody who has access to the Word document and OriginPro installed on their computer in the same way. Changes made to the embedded graph are obviously not carried across to the original OriginPro source file and only exist within the Word file.

Tables

Tables are effectively rows and columns of numbers arranged under headings and provide a useful data presentation alternative to figures for thematically similar data that should ideally appear together but is difficult to compile onto axes in a single figure due to conflicting parameters, units or scales. The ability of a single table to seamlessly present data that would otherwise need to be presented over multiple figures makes tables an especially useful tool in manuscripts that contain a lot of data but have strict figure limits, as are imposed by some journals.

That said, patterns are not well represented in tables, with figures providing a readily accessible visual interpretation of important trends that is much more intuitive for readers. Tables subsequently appear less frequently in manuscripts than figures and typically serve an archival function, presenting data that supports the manuscript message but is not the central focus. The importance and relevance of the data that a table contains should, however, not be blemished by this archival function. With tables missing the engaging visual aspect that figures enjoy, it can be easy to blur the lines between critical and less important content. Tables should only contain important and relevant information in exactly the same way as figures.

Table Formatting

Various publishers have their own formatting preferences for tables in terms of lines, fonts, and coloured fill or shading. They will implement these themselves during the production process if they choose to publish the manuscript in a journal or book. Exceptions exist for some publishers that provide a template including preformatted editable tables for the author to use (common for conference papers and applicable to some journals and books). However, authors are generally not encouraged to style their tables themselves. Tables appearing in university theses should also be free of styling.

Table formatting should be restricted to three horizontal lines (FIGURE 12). The header row of the table should be bounded by an upper and lower horizontal line and the text that it contains bolded. All other table content should use regular typeface. The final row of the table should also be bounded by a single horizontal line.

Table 1. Elemental analysis (C, H, N in wt%), ultimate tensile strength (σ_{UTS}, MPa) and elastic modulus (E, GPa) of untreated and diimide treated (hydrogenated) natural rubber balloons aged (12 or 24 months) under ambient and elevated temperatures (T_{aging})

Balloon material	Age (months)	T_{aging}	Elemental composition (wt%)			σ_{UTS}	E
			C	H	N	(MPa)	(GPa)
Natural rubber[a] (untreated)	12	Ambient	43.2 ± 0.1	6.5 ± 0.1	3.5 ± 0.1	18.3 ± 0.2[c]	0.1 ± 0.01
		Elevated[b]					
	24	Ambient					
		Elevated					
Hydrogenated natural rubber (diimide treated)	12	Ambient					
		Elevated					
	24	Ambient					
		Elevated					

[a] Balloons had small manufacturing defects which caused discrepancies in results

[b] Temperature varied due to inherent changes in the weather

[c] Specimens tore before tensile failure

FIGURE 12. *Setting up a table in Microsoft Word. Be sure to use a descriptive table caption, structure header row and stub column parameters intuitively and hierarchically, use parameter symbols, abbreviations, and acronyms in the header row where possible accompanied by associated units, provide an indication of the accuracy of table values, maintain decimal place consistency within the same column and add footnotes where appropriate.*

Open the 'Insert' tab in Microsoft Word and select 'Table'. A table of the desired dimensions can then be generated through the interactive grid or by designating the number of rows and columns for the new table from the 'Insert Table' option in the 'Table' menu. The bounding lines can then be set by selecting the relevant rows of the table in turn (top and then bottom rows), right clicking on the selected row, choosing 'Borders' and then 'Top Border' or 'Bottom Border' as appropriate.

Do not use Microsoft Word 'Table Design' tools and do not use coloured fill or shading in any of the cells. Line spacing is a matter of author preference but should typically be at least 1.5 pt. This is achieved by selecting all table cells and then choosing the desired spacing from the 'Line and Paragraph Spacing' dropdown field on the 'Home' tab. It is also important to note that a table should appear in a manuscript as plain text and not as an image.

Dependent variables are listed in the columns. Even with a relatively small number of columns and fairly short variable descriptions as column headings, it can be difficult to fit everything into the header rows of a table without the cells expanding vertically to spill over multiple lines. Try to group variables that have elements of their description in common using spanners i.e., using an additional row to house the common elements and then shorter descriptions distinguishing differences in the row below. Use symbols and abbreviations to reduce the text required in header rows where possible with a definition for the symbol in the table caption. All parameters appearing in the header row must be accompanied by their unit, often listed in brackets.

Independent variables, such as sample types, environmental or experimental conditions, are listed as row headings in the first 'stub' column. Multiple levels of hierarchy can be used across several columns, if necessary, to structure the data in a neat and intuitive format. If multiple hierarchical levels are used, try to structure them logically, starting with the broadest category first and moving to the most specific. Merge cells that fall within the same category rather than repeating labelling.

Use this structuring to improve the accessibility of the table to readers. Avoid using author generated sample codes that are not immediately obvious to the reader e.g., BLNH12A and perhaps even require reference to an earlier part of the manuscript in which the coding structure was defined to be understood. Instead use simple accurate descriptions that describe the independent variables such that the reader can understand the table without referring to any other part of the manuscript.

Column and row widths and heights can be manually adjusted if desired. Otherwise, a fast method that can often be used to improve table formatting is to select the entire table, right click the selection, choose 'AutoFit' followed by 'AutoFit to Contents'. If the manuscript is left aligned, as is the case for most journal submissions, the table can then simply be centred on the page by selecting the table followed by 'Center' on the 'Home' tab. If the manuscript is aligned using 'Justify' as is the case for many theses, then right click the selected table once more and choose 'AutoFit to Window' from the 'AutoFit' menu.

Notes can also be incorporated into a table to provide additional information where necessary. This can be useful to comment on e.g., the source of specific data, definitions, explanations, conditions or exceptions to a variable or table value. These 'footnotes' are denoted using a superscript letter, asterisk or other symbol placed after the variable(s) or value(s) in the table. The footnote accompanying the designated letter or symbol is then provided directly below the table.

Number Formatting and Appropriate Use of Decimal Places

Numbers appearing in table cells should be accompanied by an indication of their accuracy, be it an error margin, range or statistics of some other kind. If no value is available for a specific cell, then this can be indicated with a single dash (-) rather than by leaving the cell empty. Also note the difference between a measured 0 value and a lack of measured data and never confuse the two. The use of an average and standard deviation is common in many research areas e.g., 1.0 ± 0.1 *unit*, although the most appropriate method to express uncertainty will vary by manuscript and field. The AVERAGE() and STDEV() functions in Microsoft Excel are often the fastest and simplest method for students to calculate the average and standard deviation of their data if desired.

The level of accuracy of the numbers being presented should also be reflected in the number of decimal places that accompany them e.g., listing too many decimal places may sometimes misrepresent the precision of data. The number of decimal places should reflect the accuracy of the instrument. If measurements were made by eye using a ruler, then listing measured values with two decimal place accuracy is totally inappropriate. Even one decimal place accuracy is probably a stretch. The same applies to results derived from computer software which spits out numbers with multiple decimal place accuracy that are based on less precise input data.

On the other hand, some instruments provide incredibly accurate data with multiple decimal place accuracy for properties that, depending on the application, may not require the use of all decimal values. Rounding to one or two decimal places where appropriate can help to remove clutter in tables.

Regardless of how many decimal places are used, consistency should be maintained for all numbers appearing in the same column. This is also true for the standard deviation or error if relevant and for coincidentally whole numbers appearing in a column containing decimals. These can then be represented with zeros following the whole number e.g., 1.00.

Schematics

Generating attractive schematics can be a challenging task for science students. There is typically little to no provision for creative arts instruction in scientific degrees. Science students may subsequently have limited exposure to and experience in drawing and other artistic skills that could be useful in illustrating a scientific manuscript.

Fortunately, the emphasis in scientific schematics is not typically on artistic value but rather on the value that an illustration provides in explaining and visualising a scientific concept that may be challenging to express using words. There is no expectation that a scientific schematic will resemble illustrations produced by professional graphic designers, although some publishers and high impact journals may provide design services free of charge to help authors improve their schematics after their manuscript has been accepted for publication.

Generating and Combining Picture Elements

There are a number of strategies that can help to make the schematic design process easier for less artistically gifted students. Regardless of creative talent or training every person with a scientific or technical background has skills that they can exploit to improve their schematics. Engineers may have experience in computer aided design (CAD) software that they can use to generate models. Chemists and physicists may be familiar with software that can visualise chemical or atomic structures and interactions. Such software can be used to generate content for schematics.

Clip art, vectors and other public domain or open access content available on the internet or in the libraries of some word processing or drawing software can also be a good source of visual content for schematics. Try sites like Wikipedia (www.en.wikipedia.org) or Pexels (www.pexels.com) for images, and Flat Icon (www.flaticon.com) or word processing or drawing software libraries for symbols. Figures or parts thereof appearing in other open access journal articles can also be used in your schematic. If desired, subscription libraries, e.g., Adobe Stock (www.stock.adobe.com) also offer content for purchase. In some cases, e.g., images licensed under a Creative Commons license, the source of the image must be attributed in the figure caption. Check the image license for more details. Images sourced from the public domain, images for which a license is purchased and clip art provided in word processing and drawing software can usually be used without attribution, although this should be checked in the license or program.

Elements that will be used to generate a sample schematic in this guide are displayed in FIGURE 13. They include shapes generated and styled in Microsoft Word, a CAD model generated using Dassault Systèmes CATIA (www.3ds.com), clip art sourced from EDrawMax (www.edrawsoft.com) and public domain molecular schematics featured on Wikipedia.

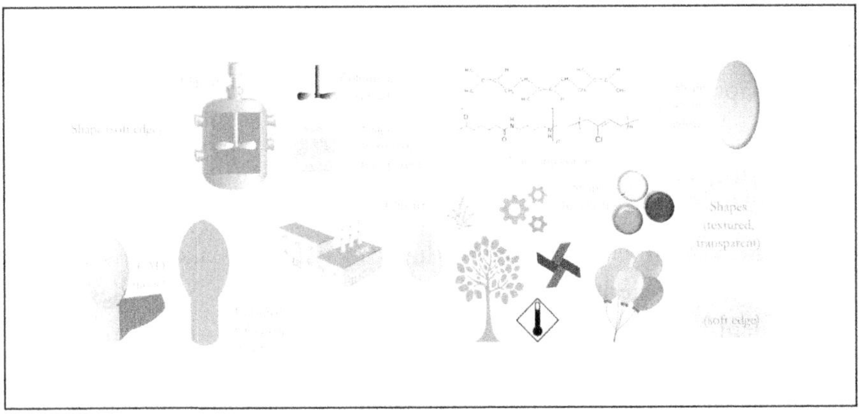

FIGURE 13. *Visual elements that will be used to produce a sample schematic on the balloon manufacturing process for this guide. Colours appearing in this figure may not be interpretable on some devices and in some print formats.*

Layering of many independently sourced elements in a single space with emphasis on attractive spacing and symmetry can produce great results irrespective of the author's artistic skill. Although graphic design software will of course yield better results, even Microsoft Word and PowerPoint offer a considerable range of creative tools and are easy to use. The following information relates to Microsoft Word.

Source the elements for your schematic from clip art, stock image libraries or online and save them in a folder on your computer as image files. Import the independent elements into Microsoft Word as pictures and remove their backgrounds using the 'Remove Background' tool on the 'Picture Format' tab as necessary. Right click each image and set its 'Wrap Text' option to 'In Front of Text' so that you can move and layer them easily without them interacting with each other. Experiment with overlaying multiple components to generate new geometries, bringing elements to the front or sending them to the back as necessary by right clicking the element and selecting the appropriate option. The 'Corrections', 'Color', 'Artistic Effects' and 'Transparency' options on the 'Picture Format' tab can also be useful in modifying and combining schematic components.

Generating and Styling Shapes

Basic shapes generated using Microsoft Word can be very useful and even look attractive when combined and formatted correctly (FIGURE 14). They are generated by selecting 'Shapes' on the 'Insert' tab and choosing the desired shape. The colour and weight of the shape's bounding lines can be modified using the 'Shape Outline' option on the 'Shape Format' tab. Here it is also possible to select from solid lines or a variety of dashed line options and whether the line appears as a hard straight line or more roughly sketched. If no line bounding the shape is desired, then the 'Shape Outline' can be set to 'No Outline'.

If a shape should contain a fill, then this can be added using the 'Shape Fill' option on the 'Shape Format' tab. If the shape should be hollow, then 'No Fill' should be selected. For shapes that do use a fill, a solid colour or 'Gradient' fill may be used but restricts the styling options to those in the 'Shape Effects' menu of the 'Shape Format' tab ('Shadow', 'Reflection', 'Glow', 'Soft Edges', 'Bevel', '3-D Rotation'). Using the 'Texture' option for the 'Shape Fill' on the 'Shape Format' tab converts the shape into a picture and allows the use of the 'Picture Format' tab in addition to the 'Shape Format' tab. The 'Picture Format' tab contains other useful tools such as 'Corrections', 'Color', 'Artistic Effects' and 'Transparency' which can provide additional flexibility during shape styling. If access to the 'Picture Format' tab is desired but a texture fill cannot be used, then a shape with a 'Solid' or 'Gradient' fill can be converted into a picture by right clicking it and selecting 'Copy' and then right clicking in a blank space within the document and selecting 'Picture' under the 'Paste Options'.

The 'Shadow' 'Reflection' and 'Bevel' options in the 'Picture Effects'

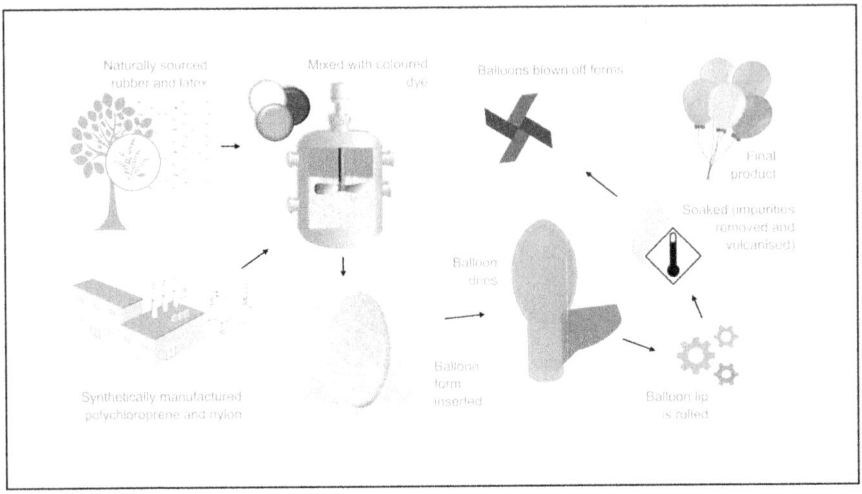

FIGURE 14. *A sample schematic detailing the balloon manufacturing process produced in Microsoft Word using a CAD model, clip art and modified shapes. Colours appearing in this figure may not be interpretable on some devices and in some print formats.*

menu of the 'Picture Format' tab in particular can help to add depth and dimension to ordinary shapes. The 'Soft Edges' tool within the same menu can also be useful when blending schematic components.

Once the final schematic has been produced, its individual elements (including all pictures, shapes, and text boxes) can be grouped and pasted into the manuscript draft as they appear, or the entire thing can be converted into an image and inserted into the draft separately. If the latter is preferred, then ensure that the Microsoft Word file used to generate the schematic is saved as a separate document in case modifications are required in the future.

Some loss of quality is typically associated with converting a Microsoft Word file to an image. However, the benefits of working with an image that can be moved and resized with ease as opposed to a group of elements which may behave unpredictably when their position or size is altered often outweigh this drawback. A schematic is most easily extracted from Microsoft Word by capturing it using the default Snipping Tool in Microsoft Windows. However, this rudimentary process will often yield poor quality images which may include unwanted elements, such as the text cursor and grammar correction suggestions. A better option is to save the Microsoft Word page as a PDF and then use a PDF reader, such as Adobe Acrobat Reader DC (free) or Pro DC (subscription) (www.acrobat.adobe.com) to export the page as an image.

If using Adobe Acrobat Reader DC, open the PDF and select 'Convert to Word, Excel or PowerPoint' from the 'File' menu. Then select 'Image' from the 'Convert To' menu, set the image format and quality and export the image. The entire page will be exported as an image and can be cropped using graphic design software, Microsoft Paint or Word. The schematic should ideally occupy most of the page space or this cropping will be associated with considerable losses in image quality.

If using Adobe Acrobat Reader Pro DC, open the PDF and crop the page to the schematic size using the 'Crop Pages' tool from the 'Edit PDF' menu. Once the page is cropped, select 'Convert To', 'Image' and then the desired image format from the 'File' menu. The desired image settings can then be configured by clicking the 'Settings' button prior to saving.

For slightly more advanced users, Microsoft PowerPoint also allows slides to be exported as high resolution (e.g., 1200 dpi) images. Simply change the export resolution in the Windows Registry Editor as documented at docs.microsoft.com/en-us/office/troubleshoot/powerpoint/change-export-slide-resolution. Then create the schematic on a PowerPoint slide, select 'Save As' and choose 'JPEG File Interchange Format' or another preferred format from the format box. This often represents a better-quality output file than that achieved using a Word and PDF export combination and is one advantage of using PowerPoint over Word for schematic design.

Figure and Table Captions

Captions are the text that accompany figures and tables, describing their content and context. They typically appear below figures and above tables. Each caption serves to support the critical free-standing function that figures and tables should serve. Even very carefully designed figures and tables will contain elements that need to be explained in words if a reader is to understand them without first reading the manuscript text.

Each caption begins with a designation of its number e.g., 'Figure 1' or 'Table 1'. Numbering is sequential throughout the manuscript and figure and table numbering are distinct. The figure or table number is then followed by a detailed description of the displayed content, context, notable features and if appropriate a brief interpretation of the key results.

Always write captions after the figure or table is already finished rather than before. Aim to describe the figure or table using as much detail as possible. Note all samples, measured data and analysis types appearing in the figure or table using a hierarchical listing style to introduce the broadest theme first, such as measured data or analysis type and parameters that are used to express them (axis labels for figures or column headings for tables), followed by the test variables and finally individual samples. This method helps to avoid repeating identical or similar words.

If letters are used to distinguish various sections of a figure or table from each other, then maintain labelling consistency in the caption. Also, define any symbols, abbreviations or acronyms that appear in the figure or table and provide their associated units and legend information, such as the shape and colour of markers, lines, or both. This can be incorporated within the sentences describing the figure or table content with the parameter, term or word written in full followed by the symbol, abbreviation, or acronym in brackets or as a group at the end of the caption in a list preceded by an appropriate description e.g., 'Acronyms:'.

EXAMPLE - FIGURE AND TABLE CAPTIONS

FIGURE 1. *Ultimate tensile strength (σ_{UTS}, MPa) and elastic modulus (E, GPa) of untreated (NR, blue solid squares) and diimide treated (hydrogenated, HNR, green hollow circles) natural rubber balloons with respect to age at elevated temperature (months). Mean values are marked with broken black lines.*

FIGURE 2. *SEM micrographs detailing the fracture surfaces of (a) untreated and (b) diimide treated (hydrogenated) natural rubber balloons exposed to elevated temperatures for 24 months. Microcracks are marked with white arrows.*

FIGURE 3. *Balloon manufacturing process. Raw materials are sourced from the environment (rubber) or synthetically manufactured (polychloroprene and nylon) and mixed with a coloured dye. Balloon forms are then dipped into this mixture to coat them and allowed to air dry. A lip is mechanically rolled onto the balloon, it is soaked to remove impurities and vulcanise before being blown from the form.*

TABLE 1. *Elemental analysis (C, H, N in wt.%), ultimate tensile strength (σ_{UTS}, MPa) and elastic modulus (E, GPa) of untreated and diimide treated (hydrogenated) natural rubber balloons aged (12 or 24 months) at ambient and elevated temperatures (T_{aging})*

Briefly highlight the presence and importance of any critical relationships, regions of interest or notable results that are marked in the figure or table. However, keep these explanations short i.e., one sentence for each point made. There should never be an aspect of a figure or table that attracts attention from the reader yet is not comprehensively described. At the same time, captions should be as short and concise as possible. Simply state the point. More extended discussion should be saved for the manuscript text.

If a figure or table contains data sourced from literature, then this should also be acknowledged in the caption. Indicate this at the end of the caption where appropriate e.g., 'Data from *reference*' or 'Compiled from literature (*reference*)'. In tables, these references can alternatively be listed in a separate column or footnotes.

Once the captions are finished, paste them with their accompanying figures and tables into a new Microsoft Word document (the manuscript draft). Rearrange them, if necessary, to achieve a logical flow. Can you understand the content and context of every figure and table from nothing more than the caption? Arranged together, do the figures and tables tell the story of the manuscript even without the words? Ask someone else who is not familiar with the content of your manuscript the same questions.

Unit Formatting

Units are exceptionally important in science. Without them numbers have no context and hence no meaning. Appropriate SI units (International System of Units) should accompany all values in the manuscript text, figures and tables and all parameters in equations. These should be expressed using abbreviations where possible.

Units should be capitalised correctly e.g., g, L, Pa, N etc., and contain a single space between the parameter or value and its unit, with the exception of units such as degrees (°) and percentages (%). Unit abbreviations are not pluralised e.g., h for hours rather than hs or hrs, m for metres rather than ms and can be expressed using a fraction or exponent style representation e.g., g/m^2 or $g\ m^{-2}$ (FIGURE 15).

$$\mathbf{g/m^2} = \frac{\text{g (numerator)}}{\text{m}^2\text{ (denominator)}} = \text{g} \cdot \frac{1}{\text{m}^2} = \mathbf{g\ m^{-2}}$$

FIGURE 15. *Variations in unit notation and the mathematical logic behind their differences.*

Appropriate unit prefixes should be used, e.g., nano- (10^{-9}), micro- (10^{-6}), milli- (10^{-3}), centi- (10^{-2}) deci- (10^{-1}), deka- (10^{1}), hecto- (10^{2}), kilo- (10^{3}), mega- (10^{6}), giga- (10^{9}) to more efficiently express very small or very large numbers e.g., 1,830,000 Pa = 1.83 MPa. Multipliers and prefixes are used in the unit numerator and not the denominator and simplification should be performed where possible e.g., MN/mm^2 can be expressed as N/m^2 which is in turn pascals (Pa).

Equations

Use of mathematical equations is an inherent aspect of most scientific work. These can range from simple routine calculations to very complicated algorithms and modelling and may be used behind the scenes during data processing or even constitute a scientific result of their own. The relevance and importance of an equation to a scientific manuscript will influence the most appropriate way to present it.

It is possible to use equations within sentences provided they are punctuated accordingly, e.g., if an equation is used at the end of a sentence, then it should be followed by a full stop. This may be appropriate in theses which contain many equations that must be documented, perhaps used sequentially in an extended calculation process. This method for incorporating equations in a scientific manuscript can, however, be very clunky. Sentences filled with equations may not read well and can appear cluttered and visually unappealing.

A better way to incorporate equations into a manuscript can be to formulate a relevant paragraph that describes and then refers to them in the same way as for figures, and tables. Equations can then be presented as a group at the end of the paragraph.

Equations that are presented on separate lines rather than in sentences should be either indented or centred and labelled with a right justified equation number in brackets e.g., (1). Equation numbering should be sequential throughout the manuscript. Every labelled equation must be referred to in the text with an appropriately formatted label e.g., (Eq. (1)). Journal articles or conference papers use a single number when labelling equations. Thesis and book chapters often indicate which chapter an equation belongs to by incorporating the chapter number within the equation number in the same way as for figures and tables e.g., the first equation in the first chapter is Eq. (1.1). The numbering sequence then starts again at the beginning of the next chapter e.g., Eq. (2.1). Even if an equation spans multiple lines, which can be the case with long equations, it is still only assigned a single equation number.

Equations should comprise symbols, abbreviations, and acronyms where possible. If words or phrases must be included, then they should appear in lower case. All parameters contained within equations must be defined within the manuscript text. Readers should not have to guess what any symbols, abbreviations or acronyms represent, regardless of how obvious their meaning may seem. Italics should be used for all parameter symbols both in the equation and in the manuscript text, except for Greek letters, for which no italic form traditionally exists. Numbers, units, and functions are also not italicised.

To insert an equation in Microsoft Word, select 'Equation' from the 'Insert' tab and use the tools in the 'Equation' tab to format the equation as desired. The default font for equations in Microsoft Word is Cambria Math in size 11 pt. The size of this font may or may not correlate well with the size of the font used in the rest of the scientific manuscript, depending on which font was selected for the main text body. Font size should generally be matched but if a considerable size difference exists between the equation and text body the equation font can be enlarged or reduced to make this difference less apparent.

It is not always necessary to list every single equation used in every process in the scientific manuscript, especially in publications such as journal articles and conference papers. Equations and calculations that are a matter of common knowledge e.g., averages and standard deviations, can be omitted. Published equations, methods and algorithms also don't need to be explicitly reproduced in a scientific manuscript unless considered absolutely necessary. Simply state what was used and provide a reference to the original publication in which it was described.

EXAMPLE - PARAMETER DEFINITION AND EQUATIONS

The balloon material grammage, *G* (Eq. (1)) was determined from the mass, *m* and area of circular 100 mm diameter specimens (r = 50 mm) using a micro-balance (*model and manufacturer*). The thickness (*d*) was measured using a digital outside micrometer (*model and manufacturer*), with 20 replicate measurements across the surface, and used to calculate the envelope density ($\rho_{envelope}$) of the material (Eq. (2)). The porosity (Φ) was then calculated from the measured skeletal ($\rho_{skeletal}$) and envelope densities (Eq. (3)).

$$G\ (\mathrm{g\ m^{-2}}) = \frac{m\ (\mathrm{g})}{\pi \cdot r^2\ (m)^2} \tag{1}$$

$$\rho_{envelope}\ (\mathrm{g\ cm^{-3}}) = \frac{G\ (\mathrm{g\ m^{-2}})}{d\ (\mathrm{\mu m})} \tag{2}$$

$$\Phi\ (\%) = \left(1 - \frac{\rho_{envelope}\ (\mathrm{g\ cm^{-3}})}{\rho_{skeletal}\ (\mathrm{g\ cm^{-3}})}\right) \cdot 100 \tag{3}$$

RESULTS AND DISCUSSION

The 'Results and Discussion' section of a manuscript is its core. Here the experimental results are interpreted by the manuscript author and presented to the reader in a logical and easy-to-digest fashion. This section constitutes the key scientific value of the manuscript and should be matter of fact, presenting the results as they are, unembellished, unblemished but always in a concise, clear, and well-explained fashion. The responsibility of the author to interpret and present the results as a polished product should be emphasised here.

The results and discussion should flow as a convincing story supported by relevant figures and tables, data interpretation and comparison, explanations, and references to other literature. There is no room for unsupported author speculation, unprocessed, unclear, or mismatched data, vague explanations, or hypotheses. Results and discussion can be presented separately, however, a combined 'Results and Discussion' section is currently common practice in most scientific documents.

Write this section based on the 'Restructured Summary of Results' and 'Data Presentation Plan' produced in WRITING AN ORIGINAL RESEARCH MANUSCRIPT — OUTLINE OF RESULTS AND DISCUSSION. It is worth carefully rescrutinising the steps completed in this 'Outline of Results and Discussion' section at this point before continuing as any errors in the scope of the outline will be carried through to the results and discussion either weakening it or compromising it completely.

- Have all variables and measured data been considered for inclusion? Is the 'Summary of Results' complete and are the claims fully supported by the measured data?
- Is the 'Central Message' short, sharp and clear? Does it accurately reflect the results?
- Are the results grouped logically in the 'Restructured Summary of Results' and do these grouped points flow logically as a story with a central theme?
- Does the 'Data Presentation Plan' efficiently and logically categorise results and data into groups that can be presented together?
- Have these groups been screened for their importance to the story and 'central message' of the manuscript with less important groups pegged for appearance as supplementary material or appendices rather than content in the main body of the manuscript?
- Has the appropriate medium of expression been identified for each group? e.g., comment in-text with values, graph, schematic or table.
- Have figures and tables been generated based on the instructions in WRITING AN ORIGINAL RESEARCH MANUSCRIPT — FIGURE AND TABLE DESIGN.
- Have these figures and tables been inserted into a Microsoft Word document and arranged by theme to flow and deliver the intended manuscript story i.e., structured based on the 'Restructured Summary of Results'?

If the answer to all these questions is yes, then it is now time to write the 'Results and Discussion' section of the manuscript. Open the Microsoft Word document containing the figures, tables and their associated captions arranged by theme and flowing to deliver the intended manuscript story (the manuscript draft). Paste the relevant bullet point(s) associated with each figure or table from the 'Restructured Summary of Results' directly below each of them. These bullet points comprise the main message that each figure and table should embody and must shine through in the corresponding results and discussion of each in the text.

Now look at each figure and table in turn and formulate an interpretation of the key trends and results within it that support its message as a separate list of bullet points. Which aspect of the manuscript's 'central message' and story are you attempting to convey with the figure or table's inclusion and which of its attributes best illustrate this?

Examine all important features of each figure and table. If variations in a property across various samples are detailed, perhaps there are samples that performed particularly well or badly. Perhaps there are samples that did not perform as expected or exhibited behaviour or properties that varied radically from the other samples. If a variable is being tested, then there may be trends in the results. Summarise the most meaningful aspects of the figure or table to the manuscript story in the bullet point list.

The assembly of a writing plan based on the produced figures, associated relevant bullet point(s) (main message) from the 'Restructured Summary of Results' and meaningful aspects of the figure that support this main message is illustrated in Figures 16, 17 and 18.

EXAMPLE - RESULTS AND DISCUSSION WRITING PLAN FOR FIGURES

Relevant point(s) from the 'Restructured Summary of Results' – the message to be conveyed:

- Sample characterisation: *samples* (e.g., diimide treated balloons) had a higher *measurement* (e.g., percentage hydrogenation) than the others.

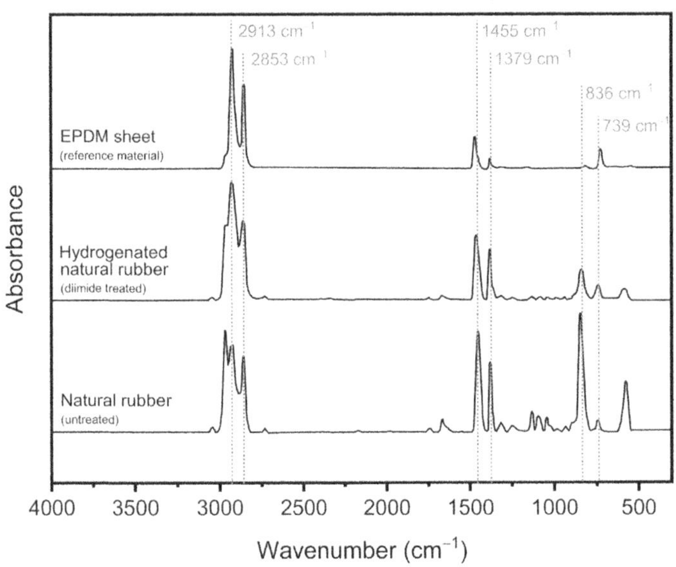

FIGURE 16. *Fourier-transform infrared (FT-IR) spectra for untreated and diimide treated (hydrogenated) natural rubber balloon material and a commercial ethylene propylene diene monomer rubber (EPDM) sheet reference.*

Important features in the figure that convey this message and will be used as inspiration for the results and discussion text:

- Untreated compared to diimide treated natural rubber: diimide treatment causes increased hydrogenation – decrease in band intensities at 836 cm^{-1} (C–H bending), 1379 cm^{-1} ($-CH_3$ bending), 1455 cm^{-1}, ($-CH_2-$ scissoring), 2853 cm^{-1} ($-CH_3$ symmetric stretching). Increase in band intensities at 739 cm^{-1} ($-CH_2-$ rocking) and 2913 cm^{-1} ($-CH_2-$ asymmetric stretching).
- Diimide treated (hydrogenated) natural rubber and the EPDM sheet reference material have similar molecular structure (bands at similar wavenumbers).

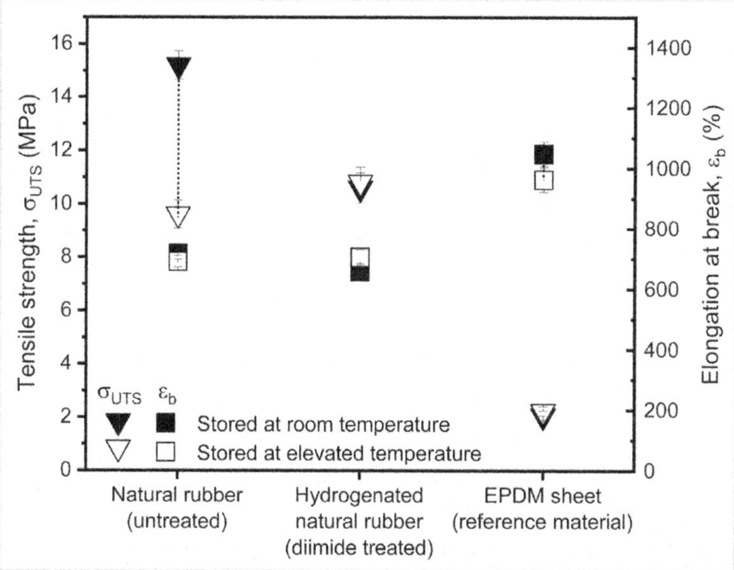

FIGURE 17. *Tensile strength (σ_{UTS}, MPa) and elongation at break (ε_b, %) of untreated and diimide treated (hydrogenated) natural rubber balloon material stored for 24 months at ambient and elevated temperature and a commercial EPDM sheet reference.*

Relevant point(s) from the 'Restructured Summary of Results' – the message to be conveyed:

- No *samples* (e.g., balloons) stored at *condition* (e.g., ambient temperature) exhibited mechanical property degradation over the *test period* (e.g., 24 months). Morphological analysis [omitted from this example] showed *observation* (e.g., no cracking) in these samples. Conversely, all *samples* (e.g., balloons) stored at *condition* (e.g., elevated temperatures) exhibited *observation* (e.g., mechanical property degradation) except for those treated with *compound* (e.g., diimide). Morphological analysis also showed *observation* (e.g., cracking) in all *samples* (e.g., balloons) stored at *condition* (e.g., elevated temperature) except those treated with *compound* (e.g., diimide).

Important features in the figure that convey this message and will be used as inspiration for the results and discussion text (example for the effect of storage temperature on the tensile properties of untreated and diimide treated (hydrogenated) natural rubber balloon materials):

- Untreated natural rubber balloon material is susceptible to a ~37% reduction in tensile strength when thermally aged (15.2 MPa at ambient temperature to 9.6 MPa at elevated temperature), although its elongation at break is largely unaffected.
- Untreated compared to diimide treated (hydrogenated) natural rubber balloon material: the diimide treatment results in a ~29% lower pre-thermal aging tensile strength (10.8 MPa compared to 15.2 MPa at ambient temperature) but tensile strength is virtually unaffected

by thermal aging (elevated temperature) in diimide treated balloons. The elongation at break of both materials is similar (659% and 717% with and without diimide treatment, respectively) and the elongation at break of diimide treated material is not affected by thermal aging.

- Untreated and diimide treated (hydrogenated) natural rubber balloon materials are at least 5 times stronger than the reference material both pre- and post-thermal aging. However, the EPDM reference has an elongation at break 46-59% higher than the two balloon materials.

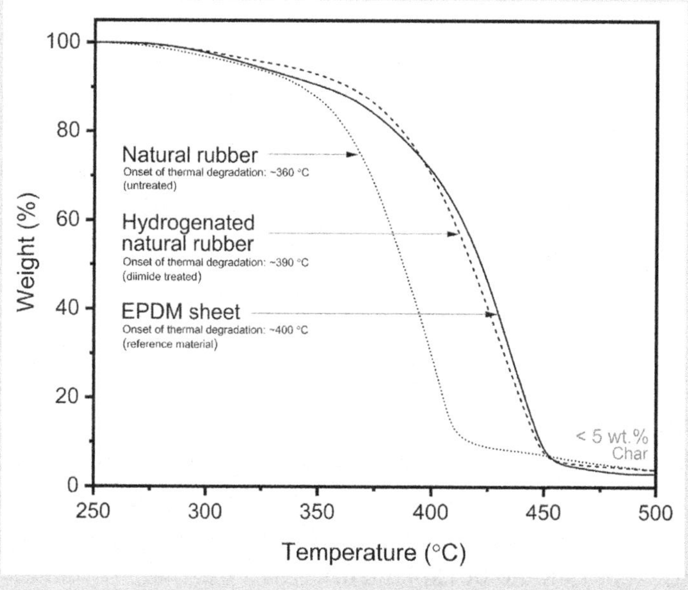

FIGURE 18. *Thermogravimetric analysis (TGA) of untreated and diimide treated (hydrogenated) natural rubber balloon material and a commercial EPDM sheet reference.*

Relevant point(s) from the 'Restructured Summary of Results' – the message to be conveyed:

- Thermal analysis indicated that *samples* (e.g., diimide treated balloons) exhibited greater resistance to thermal degradation than the others. We attribute this to the treatment of the material with *compound* (e.g., diimide). This is supported by theory, literature, or further experiments.

Important features in the figure that convey this message and will be used as inspiration for the results and discussion text:

- Diimide treated (hydrogenated) natural rubber balloon material has an onset temperature of thermal degradation ~30 °C higher than the untreated variant.
- All materials have approximately the same char content (< 5 wt.%).
- The thermal degradation properties of diimide treated (hydrogenated) natural rubber balloon material resemble those of the reference material.

Attributes of the figure or table that do not relate to its message and place in the manuscript story should receive very little emphasis or even be omitted from the discussion completely. An exception to this exists if samples or trends were identified as important in an earlier figure or table, in which case the discrepancy in their relevance later in the manuscript (i.e., lack thereof) should be explained.

Once bullet points that embody the essence of each figure or table have been produced, read through them to ensure that together they flow as a story to convincingly deliver the manuscript's 'central message'. If they do then the bullet points can be further developed into connected prose using the suggestions provided on 'Internal Comparison' and 'External Comparison and Discussion' below.

Split results and discussion content into relevant themed subsections each with a unique heading. Headings should use full sentences that provide all necessary experimental context e.g., 'physical and mechanical properties of the balloons' rather than simply 'physical and mechanical properties'. Once all figures and tables have been discussed, the manuscript 'Conclusion' can be written. This process is covered in the 'Conclusion' section of this guide.

Internal Comparison

Internal comparison focusses on interpreting and extracting meaning from data. It is important to comment on the spread of the data, trends and notable values, samples, or properties, how they compare with each other and with the reference if applicable. High or low performing samples, properties or values should be highlighted and discussed as should outlier values or unexpected behaviour. During this process it is important to remember that science should be as objective as possible. Every word, every decision, even the way in which words are ordered is open to question and criticism. Be very careful that every sentence accurately reflects the intended claims and is based on solid data. The same applies to data interpretation. Always provide an overview of all relevant data. If the overview covers every point, nobody can claim that it is biased.

That said, it is not necessary to comment on every sample, property, or value, only those which exhibit notable differences and are relevant to the manuscript story. Note the difference between a fair analysis which at least acknowledges the presence of every point with broad comments on groups of points and trends supplemented by detailed comments on notable points and an analysis that goes into excessive detail on every point. When choosing examples to discuss in detail, ensure that they are chosen for a reason e.g., they are especially relevant, best or worst performing samples, outliers. The interpretation and discussion must be objective, meaning that it is unacceptable to choose specific examples for extended discussion without having a reason to do so. Such behaviour could easily be misinterpreted as bias.

Describing Data

Data and change in data can be compared and the differences between notable values, samples, or properties described using descriptors, factors, multiples, fractions, or percentages (TABLE 2). Mathematic relationships, such as linear, exponential, and logarithmic functions, can also be used to describe change in data based on a variable. However, if mathematic functions are used to describe data sets, they should be accompanied by statistics indicating how well the data fits the function e.g., R^2 values for linear fit.

When comparing data, it is critical to do so using fair comparisons. Compare apples with apples and oranges with oranges, as the popular saying goes. It is for example not fair to compare the material properties of two radically different materials which have very different applications and functions. Comparing the tensile strength and modulus of rubber and steel for a structural application is not reasonable.

TABLE 2. *Basic but effective words that can help change the way that students think about describing data. A compulsively quantitative mindset can dramatically improve the clarity and accuracy of language in scientific writing.*

Type	*Comparison of data*	*Description of change in data*
Descriptors	Higher, lower	Increased, decreased, constant
Factors	Higher/lower by a factor	Increased/decreased by a factor of
Multiples	Double, triple	Doubled, tripled
Fractions	Was a fraction of	Increased/decreased by a fraction of
Percentages	Was a percentage of	Increased/decreased by a percentage of

Begin the internal comparison by identifying and discussing trends between variables in the study and measured data. Focus on the shape of clusters of points or the orientation of lines to assess possible trends and describe them using language such as that introduced above. Combine discussion of trends with commentary on high or low performing samples, properties, or values and quantitatively illustrate the difference between them. If reference samples have been included in the study, then these always warrant inclusion and discussion as key samples. The discussion of key trends, samples, properties, and values provides a comprehensive but focussed description of the range and shape of data distributions without discussing every data point in detail.

EXAMPLE - COMPARISON OF TRENDS BETWEEN VARIABLES AND SAMPLES

Property (e.g., onset temperature of thermal degradation) increased almost linearly with *variable* (e.g., diimide induced percentage of hydrogenation) for the *samples* (e.g., balloons), increasing by up to *value* (e.g., 100 °C) to a maximum of *value* (e.g., 450 °C).

Samples (e.g., balloons 1 and 2) exhibited the highest *property* (*value and unit*) (e.g., tensile strength), a value *percentage* higher than that of *sample* (*value and unit*) (e.g., balloon 3) and approximately double that of the reference (*value and unit*) (e.g., commercial sheet of EPDM).

When identifying key data points, it is a good idea to provide their associated numerical values and units in brackets directly after they are first mentioned in text. The same is true for change in data based on experimental variables, where initial and final numerical values can be compared in brackets. This makes comparison of important points and changes in the data very quick and easy for the reader and reduces their reliance on the figure, from which the extraction of highly accurate information can sometimes be time consuming or challenging.

EXAMPLE - INCLUSION OF VALUES WITHIN WRITING

Property (e.g., onset temperature of thermal degradation) increased with *variable* (e.g., diimide induced percentage of hydrogenation) for *samples* (e.g., balloons 1, 2 and 3), increasing by a factor of *value* for *sample* (*value and unit* to *value and unit)* (e.g., balloon 1)*, value* for *sample* (*value and unit* to *value and unit)* (e.g., balloon 2) and *value* for *sample* (*value and unit* to *value and unit*) (e.g., balloon 3), compared to the reference (*value and unit*) (e.g., commercial sheet of EPDM).

Improving Communication of Data

It is sometimes necessary to list many key data points in text and in such cases the flow of the sentence can suffer. Sentence reordering using the word 'respectively' can help to reduce this problem. Appearance of the word 'respectively' in a sentence indicates that the order in which numbers are listed in the sentence corresponds with the order in which their associated labels will be listed later in the same sentence and vice versa.

EXAMPLE - USING 'RESPECTIVELY' TO REORDER SENTENCES

Property (e.g., onset temperature of thermal degradation) increased with *variable* (e.g., diimide induced percentage of hydrogenation) for *samples* (e.g., balloons 1, 2 and 3), increasing by factors *value*, *value* and *value* for *sample* (*value and unit* to *value and unit)* (e.g., balloon 1), *sample* (*value and unit* to *value and unit)* (e.g., balloon 2) and *sample* (*value and unit* to *value and unit*) (e.g., balloon 3), respectively, compared to *reference* (*value and unit*) (e.g., commercial sheet of EPDM).

Samples that exhibit similar performance, properties or values can be grouped and commented on using a range. This is true of notable samples, properties and values that fall in the high and low ranges in addition to unremarkable data that falls in between, although such data often does not warrant considerable discussion or even mention in some cases.

EXAMPLE - COMMENTING ON DATA USING RANGES

Samples (e.g., balloons 1, 2 and 3) typically exhibited higher *property* (*range*) (e.g., onset temperature of thermal degradation) than *samples* (*range*) (e.g., balloons 4, 5 and 6).

When discussing figures in text it is often most appropriate to discuss by property. This is because figures often represent measured data from a single instrument or analysis due to the need for a common axis for plotting. However, in some cases it can be more efficient or effective to discuss by sample. A good example of this is when it is desired to compare two very different samples across multiple properties displayed on a graph with multiple vertical axes or if a table containing many parameters is to be discussed. Nonetheless, this kind of comparison will quickly become very challenging if the number of samples compared is increased.

EXAMPLE - DISCUSSION BY SAMPLE

Sample (e.g., balloon 1) had a *property* (e.g., onset temperature of thermal degradation) of *value and unit* and a *property* (e.g., char content) of *value and unit* making it *characteristic* (e.g., high thermo-oxidative stability). Conversely, *sample* (e.g., balloon 2) had a lower *property* (*value and unit*) (e.g., onset temperature of thermal degradation) and less than half the *property* (*value and unit*) (e.g., char content). Notably, both *samples* (e.g., balloons 1 and 2) had comparable or higher *property* (e.g., first derivative of the weight loss curve) than *reference* (*value and unit* compared to *value and unit*) (e.g., commercial sheet of EPDM).

External Comparison and Discussion

The identification and discussion of outliers or unexpected behaviour signifies the end of internal comparison and the beginning of external comparison and discussion. This is because the most effective way to deal with outliers or unexpected behaviour in results is to explain why they may exist and comment on whether other studies have experienced similar abnormalities. Even if results show no outliers or unexpected behaviour, external comparison should still be included in manuscript text, comparing the results of the study with those of others and explaining the phenomena behind key results. This means finding and citing other relevant literature.

EXAMPLE - EXTERNAL COMPARISON AND DISCUSSION

Analysis (e.g., Fourier-transform infrared spectroscopy) of *sample* (e.g., balloon 1) resembled literature spectra for *material* (e.g., latex) treated with *compound* (e.g., diimide) [*reference*].

The origin of the *contaminant* (e.g., calcium carbonate) content in *samples* (e.g., balloons) is unclear but is mirrored in the commercial reference (e.g., commercial sheet of EPDM) and several other recent studies [*reference*].

Sample (e.g., balloon 2) did not exhibit *property* (e.g., rubber-like tensile modulus) instead resembling *sample type* (e.g., nylon) [*reference*].

The high *property* (e.g., tensile modulus) of the *sample* (*value and unit*) (e.g., balloon 3) suggests that it is *characteristic* (e.g., brittle), an observation shared by other recent studies [*reference*].

This indicated a considerably higher *property* (e.g., onset temperature of thermal degradation) in *sample* (e.g., balloon 1) than in *sample* (e.g., balloon 2) and *reference* (e.g., commercial sheet of EPDM) and can be attributed to *reason* (e.g., diimide induced hydrogenation) [*reference*].

Analysis (e.g., thermogravimetric analysis) also indicated that *sample* (e.g., balloon 4) exhibited high *property* (e.g., char content) and that a considerable amount of *contaminant* (e.g., calcium carbonate) was present, likely attributable to *reason* (e.g., addition as a low-cost filler) [*reference*].

Internal comparison and external comparison and discussion are interwoven as appropriate to provide a complete interpretation of the results of the study. Start sentences with their most important aspect and signpost what the reader should be looking for in the sentences that follow. If samples differed in some ways and were the same in others, then lead with the differences. They are generally more important. If a single explanation will be made over several sentences and refer to comparisons between several samples then lead with the explanation followed by sentences containing the necessary experimental proof. The reader then knows what to look for in the following sentences and can interpret the results in the context of the explanation. This is much preferred to leading with the values followed by the explanation since readers may be initially lost when stumbling onto the data without its relevant context and then have to double back several sentences later to revisit and understand it after having finally read the explanation.

Regardless of what information is being presented ensure that it is fully interpreted. Readers should be provided with an expert interpretation of the results. Not just the results. Do not simply summarise results in sentences sample by sample. Compare and discuss samples across key properties and ensure that the context and relevance of the comparisons and explanations are apparent. The reader should be implicitly aware of what is being compared, why it is being compared and how the comparison fits into the bigger picture and story (even if it only plays a very small role). Measured values should simply validate the interpretation that the reader is provided with – prove to them that the explanation they are receiving is supported by experimental evidence. Try removing the values from your writing. Do the sentences still accurately convey what happened? If not, then you might be relying too heavily on the reader's skill (or lack thereof) to interpret the numbers themselves and your explanations may require additional depth.

RESULTS AND DISCUSSION EXAMPLE

Sample (e.g., balloon 1) and *sample* (e.g., balloon 2) contained considerably less *contaminant* (*value and unit* and *value and unit*, respectively) (e.g., calcium carbonate) and *contaminant* (*value and unit* and *value and unit*, respectively) (e.g., cassava flour) LOW than *sample* (e.g., balloon 3) (*value and unit* of *contaminant* (e.g., calcium carbonate) and *value and unit* of *contaminant* (e.g., cassava flour)), HIGH which closely matched in composition that of the commercial reference (e.g., sheet of EPDM) (*value and unit* of *contaminant* (e.g., calcium carbonate) and *value and unit* of *contaminant* (e.g., cassava flour)) REFERENCE. This indicated a considerably higher *compound* (e.g., latex) content in *sample* (e.g., balloon 1) and *sample* (e.g., balloon 2) than in *sample* (e.g., balloon 3) and the reference material and can be attributed to *reason* (e.g., balloons 1 and 2 were manufactured with higher purity latex) EXPLANATION. Both *measurement* (e.g., elemental analysis) and *measurement* (e,g., char residue) of *sample* (e.g., balloon 3) resemble literature values for *contaminant* (e.g., calcium carbonate) in *literature sample* (e.g., rubber gloves) [*reference*] EXTERNAL COMPARISON. The origin of *compound* (e.g., cassava flour) in *sample* (e.g., balloon 3) is unclear OUTLIER but is mirrored in the commercial reference REFERENCE.

Scientific Writing Tips

If something is worth saying – then make it black and white

Scientific results are facts. There are usually numbers and some kind of scientific methodology behind every experiment. Avoid softening sentences with words like 'quite', 'almost', 'fairly', 'somewhat', 'should' 'could' 'maybe'. When presenting results do so in black and white terms. There should be no grey area. They are what they are. Use a confident tone and strong polarised language that conveys the strength of conviction and stark nature of results accurately. It is important to note that the message here is not to oversell author confidence in questionable results but rather to choose meaningful results to discuss that justify this confidence.

Meaningful results typically have strong characteristics or traits that distinguish them from the other data. Do not dedicate manuscript text to results that do not have a clear message, are ambiguous, potentially incorrect or can't be explained. If a result cannot be described using strong language, then consider whether it really is a meaningful result and warrants discussion as such. Leave discrepancies or uncertainties for sentences dedicated to discussion.

EXAMPLE - SPECIFIC AND POLARISED LANGUAGE

Bad: *Sample* (e.g., balloon) seemed to have a higher *property* (e.g., onset temperature of thermal degradation) than the others.

Better: *Sample* (e.g., balloon 1) exhibited a *property* (e.g., onset temperature of thermal degradation) at least 22% higher than *sample* (*value and unit* compared to *value and unit*) (e.g., balloon 2).

After the facts have been presented, inconsistencies can be discussed. Measurement error, outliers and unexpected phenomena are all an inherent part of scientific work. Don't let them affect the confidence of the manuscript tone. When dealing with them call a spade a spade. If there is insufficient data to draw any conclusions, then state as much. If it is unclear why something happened, then say so rather than coming up with a weak explanation littered with phrases like 'it seemed', 'it could be', 'it might be' or 'perhaps'. This lack of confidence will cause the audience to question author competence and the accuracy of the entire study. If authors do not have the confidence to at least use the word 'probably', then it is better to completely omit the explanation.

EXAMPLE - ACKNOWLEDGEMENT OF UNEXPECTED RESULTS

Samples (e.g., balloons 1 and 2) exhibited a *property* (e.g., onset temperature of thermal degradation) at least 22% higher than *sample* (*value and unit* compared to *value and unit*) (e.g., balloon 3). *Sample* (e.g., balloon 4) displayed considerable variation in *property* (e.g., onset temperature of thermal degradation) during testing of multiple identical specimens cut from the same balloon and could not be compared with the other *samples* (e.g., balloons). This variation is probably attributable to *explanation* (e.g., inconsistencies in latex composition).

Be compulsively concise and precise

Scientific writing should be as precise and concise as possible. Do not use unnecessary words or phrases. Use blunt and direct sentences instead. For example, no sentence need start with "The results suggest that". These are useless and wasted words. State the results without ado: "Balloons treated with *compound* exhibited higher tensile strengths at elevated temperatures than those which were not".

This also applies to wordy sentences. Do not use multiple words when one would suffice e.g., 'agree' rather than 'are of the same opinion' and 'conversely' instead of 'on the other hand'.

EXAMPLE - MINIMISATION OF WORD COUNT

Sample (e.g., balloon 1) treated with *compound* (e.g., diimide) exhibited *property* (e.g., onset temperature of thermal degradation) of *value and unit*. Conversely, *sample* (e.g., balloon 2) lacking treatment with *compound* (e.g., diimide) exhibited a *property* (e.g., onset temperature of thermal degradation) *percentage* lower, with values up to *value and unit*.

Avoid using vague words, such as 'performance', 'problem', 'situation'. Use clear and accurate terminology that pinpoints the desired meaning with every single word. Negatives can also often be better expressed using more direct language e.g., did not succeed – failed, lacked confidence – distrusted.

Group words in sentences where possible, to avoid unnecessary repetition e.g., morphological, chemical, and mechanical properties; red, green, and blue balloons; inner and outer diameters; yield and ultimate tensile strength; tensile strength and modulus; mono-, di- and polysaccharides.

Understand the meaning of words and use them appropriately. Many words in science cannot be used as loosely as they would typically be used in English. For example, 'significantly' can only be used if the results being discussed have been proven to be statistically significant. Words such as 'considerably' are used in cases where statistical significance has not been established.

EXAMPLE - APPROPRIATE USE OF 'SIGNIFICANTLY'

The *property* (e.g., onset temperature of thermal degradation) increased significantly with *variable* (e.g., diimide induced percentage of hydrogenation) – incorrect unless this is statistically true.

The *property* (e.g., onset temperature of thermal degradation) increased considerably with *variable* (e.g., diimide induced percentage of hydrogenation) – more generic expression.

Quantify at every opportunity. The English language is very inefficient and imprecise in many aspects. Replace words such as 'some', 'several', 'many', 'most' with numbers. If comparisons are made or parameters increase, or decrease be sure to indicate by how much and provide the values. Always clearly indicate which samples, properties and parameters are being discussed.

EXAMPLE - COMPULSIVE QUANTIFICATION

Sample (e.g., balloon 1) had a 22% higher *property* (e.g., onset temperature of thermal degradation) than *sample* (e.g., balloon 2) (*value and unit* compared to *value and unit*).

Do not use words when symbols would suffice e.g., ~ approximately (for use with numbers except in cases where its meaning could be confused e.g., logic), < less than, > greater than.

Define symbols, abbreviations, or acronyms upon their first appearance in the abstract, main text body and first figure or table caption and then use the shortened version instead of their associated words in all subsequent usages. This is also true of organisms' names, which should be written in full upon first appearance in the text and shortened to the first letter of the genus followed by the species name in the rest of the text. All organism names, both full and abbreviated, are italicised.

EXAMPLE - WORD DEFINITION ON FIRST USE AND SUBSEQUENT ABBREVIATION

Genus (e.g., *Mucor*) *species* (e.g., *racemous*) fungal growth was identified in *samples* (e.g., balloons) that had been exposed to the elements for durations > *period* using *analysis* (e.g., scanning electron microscopy (SEM)). *M. racemous* hyphal filaments had an average diameter Ø of *value and unit*. Statistical analysis of SEM micrographs (p > *value*) revealed a correlation between Ø and hyphal length, with hyphae of Ø > *value and unit* typically > *value and unit* in length.

Use appropriate unit abbreviations e.g., percentage by weight – wt.% and be consistent with the use of units and abbreviations throughout the entire manuscript. This also applies to unit prefixes e.g., if most samples have a property most efficiently described using a mega prefix, then perform all comparisons with a mega prefix even if this prefix would not necessarily be most appropriate for some outlier samples. Ensure that the correct symbols are selected from the 'Symbol' tool on the 'Insert' tab in Microsoft Word: x is a letter, × is a multiplication sign, - is a hyphen, – is a minus sign e.g., 1×10^{-2} and not $1x10^{-2}$.

Every sentence in the manuscript should be supported by experimental data or analysis e.g., a figure or table, or a reference to a previously published study in the literature. Refer to figures or tables in the manuscript text e.g., (Figure 1) in the first sentence that relates to content displayed in the figure or table in question. All figures or tables that are included in a manuscript must directly relate to its content and be referred to in its text. If not, they can instead appear alongside the manuscript in the supplementary material or appendices, which can in turn be referred to e.g., (Figure S1, Supplementary Material or Appendix A). Once a figure or table has been referred to it is not necessary to explicitly refer to it again unless desired. Place figures and tables directly after the paragraph in which they are first mentioned. If a sentence is not supported by the experimental or analytical work depicted in the figures and tables in the manuscript then it requires a literature reference (see WRITING AN ORIGINAL RESEARCH MANUSCRIPT — REFERENCES), unless it appears in an abstract or conclusion section where references are not typically required.

Use of personal pronouns, such as 'I' and 'we' was historically quite common in scientific writing, as was use of the active voice (subject, verb, object). However, these conventions fell from favour as they were seen to infringe upon the 'objective' nature of scientific texts. Current practice is a matter of author preference, but personal pronouns are now generally used sparingly in scientific manuscripts and in many cases active voice is in fact preferred as it offers more direct, less confusing sentences and additional subject related information in relevant cases.

It is also generally better to begin sentences that would otherwise use the word 'it' with a proper subject, even if this clarity comes at the cost of the occasional repetition.

CONCLUSION

The 'Conclusion' is an exceptionally important part of any scientific manuscript since it essentially assesses the degree of success of the study, emphasises the key results, describes the value and impact of these results on science and society and provides the opportunity for author comment on future work, ideas, and their potential. Readers who have established their potential interest in a manuscript from the abstract will often use the conclusion and figures to assess whether the remainder of the manuscript warrants their attention.

Conclusions should focus on highlighting key results. It is important to note that key results potentially constitute only a portion of all results presented in the manuscript. Sometimes data must be included in a manuscript to support or provide context for other results. Such data does not need to be mentioned in the conclusion. Only results that relate to the 'central message' of the manuscript should appear in the conclusion.

Note: No new results and discussion or other new information should appear in the conclusion. Presentation and development of all ideas should occur in the 'Results and Discussion' section instead. Inclusion of references in the conclusion should also be avoided.

The 'Central Message' and 'Restructured Summary of Results' generated in the WRITING AN ORIGINAL RESEARCH MANUSCRIPT — OUTLINE OF RESULTS AND DISCUSSION section of this guide can help to identify the key results that should be included in the conclusion. Paste these sections into a new Microsoft Word document to work on the conclusion separately before moving it across to the manuscript draft.

Central Message

Use the 'central message' of the manuscript to formulate the first sentence of the conclusion. This sentence should accurately and concisely describe the main conclusion of the manuscript and set the scene for the more detailed key findings that will be described in the sentences which follow.

EXAMPLE - CENTRAL MESSAGE

Samples (e.g., balloons) that are typically *observation* (e.g., thermally and mechanically stable) for at least *period* (e.g., 24 months), are *observation* (e.g., more rapidly degraded) at *condition* (e.g., elevated temperatures) unless they are treated with *compound* (e.g., diimide).

Key Results

The key results of the manuscript will then be sourced from the 'Restructured Summary of Results'. However, not all results presented in this summary should necessarily be included in the conclusion. Only include results which directly relate to the 'central message' of the manuscript. Which themes does the 'central message' contain? Which experimental methods and data best illustrate the message?

EXAMPLE - RESTRUCTURED SUMMARY OF RESULTS

- Sample characterisation: *samples* (e.g., diimide treated balloons) had a higher *measurement* (e.g., percentage hydrogenation) than the others.
- No *samples* (e.g., balloons) stored at *condition* (e.g., ambient temperature) exhibited mechanical property degradation over the *test period* (e.g., 24 months). Morphological analysis showed *observation* (e.g., no cracking) in these samples. Conversely, all *samples* (e.g., balloons) stored at *condition* (e.g., elevated temperatures) exhibited *observation* (e.g., mechanical property degradation) except for those treated with *compound* (e.g., diimide). Morphological analysis also showed *observation* (e.g., cracking) in all *samples* (e.g., balloons) stored at *condition* (e.g., elevated temperature) except those treated with *compound* (e.g., diimide).
- Thermal analysis indicated that *samples* (e.g., diimide treated balloons) exhibited greater resistance to thermal degradation than the others. We attribute this to the treatment of the material with *compound* (e.g., diimide). This is supported by theory, literature, or further experiments.

Balloon properties deteriorating at elevated temperatures unless balloon material is treated with a certain compound is best illustrated with comment on the mechanical tests at a range of temperatures and compositions in conjunction with brief notes on cracking witnessed during morphological analysis (illustrates the failure mechanism behind the mechanical tests). Detailed results relating to the balloon characterisation e.g., Fourier-transform infrared spectroscopy (FT-IR) spectra and thermal analysis e.g., onset temperature of thermal degradation and char content supplement this data in the manuscript but do not illustrate the 'central message' and can be omitted from the conclusion. Only refer to this content in the context of explaining the relevant results (e.g., mechanical properties). This keeps the conclusion more focussed and improves the accessibility and clarity of the 'central message'.

When describing the key results, refrain from making direct comparisons between or commenting on specific example samples, properties or variables in the way that you did in the 'Results and Discussion' section unless their identification is important to the 'central message'. Instead try to focus on the big picture and rather describe these sample-, property- or parameter-based trends, correlations or relationships in a broader context to define the main results that support the conclusion and 'central message' of the manuscript.

EXAMPLE - CONCLUSION 'KEY RESULTS' SENTENCES

No *samples* (e.g., balloons) exhibited mechanical property degradation over the *test period* (e.g., 24 months) when stored at *condition* (e.g., ambient temperature) with *analysis* (e.g., scanning electron microscopy) indicating *observation* (e.g., no cracking) in these samples. Conversely, the *properties* (e.g., mechanical properties) of *samples* (e.g., balloons) exposed to *condition* (e.g., higher temperatures) rapidly deteriorated, with losses in *property* (e.g., tensile strength) and *observation* (e.g., cracking visible) during *analysis* (e.g., scanning electron microscopy). *Samples* (e.g., balloons) produced from *material* (e.g., latex) treated using *compound* (e.g., diimide) proved an exception to this rule, retaining their *properties* (e.g., tensile strength) even when stored at higher temperatures. The resilience of these *samples* (e.g., balloons) was attributed to *characteristic* (e.g., hydrogenation) resulting from *compound* (e.g., diimide) treatment, which *explanation* (e.g., increased the degree of saturation in the polymer and improved its resistance to heat).

Manuscript Value and Impact

With the 'central message' and key results of the manuscript now described in the conclusion, some context should be provided to their value and impact on science and society. Scientific inquiry often has a narrow field-specific scope, and the full value of results may only be obvious to other researchers working on the same topic. Even readers with broad knowledge in the research area may miss the finer details of the potential impact of a finding unless they are highlighted, let alone those less attuned to the manuscript theme. Ensure that the position of the research within a broader context and its greater purpose is explicitly noted. There should be some kind of useful application for the research, be it academic, industrial, economic, environmental, or social. Otherwise, readers may struggle to understand the sense in further investment of time and resources in the research area.

Why do your results matter? What can they be used for? Who or what could they benefit? What new opportunities or ideas do they generate? Formulate a sentence that describes the value and impact of the manuscript. For example, if consumers buy balloons in bulk and store them improperly for long periods, the balloons may well be too degraded to use by the time they are needed again. When they are subsequently thrown away, there is both a loss for the consumer and a contribution to global waste. If a compound can be used to treat the balloon material and extend its lifespan and this compound is not prohibitively expensive or associated with environmental problems, this could have ramifications for both consumers and the environment. Across the hundreds of millions of balloons sold globally every year this could represent substantial scientific and social value and impact.

> **EXAMPLE - CONCLUSION 'MANUSCRIPT VALUE AND IMPACT' SENTENCES**
>
> *Samples* (e.g., balloons) treated with *compound* (e.g., diimide) would *benefit and impact* (e.g., experience longer life spans, reducing waste and economic loss across an industry littered with bulk buying practices and discarded products no longer fit for purpose).

Author Comment and Future Perspective

There is generally limited room for unsubstantiated author opinion or hypothesis in original research manuscripts. Author opinion is more common in other manuscript types, such as Perspective articles. However, it is typically acceptable for authors to provide some final comments and perspective on future research that may follow from their manuscript if they wish. This could include ideas, concerns, suggestions, or hypotheses on the next steps for research continuing on from their study. For example, it may be worth postulating that superior compounds or additives may exist to increase balloon life and that future research should focus on identifying them. Or perhaps that environmental factors other than temperature could play a greater role in the deterioration of mechanical properties in balloons. Most relevantly in this case, however, it would certainly be worth noting that the environmental benefit associated with consumers being able to store and use more robust balloons for longer periods (waste reduction) could easily be offset by increased energy or chemical resource requirements during manufacturing or reduced biodegradability of the product at the end of its service life.

> **EXAMPLE - CONCLUSION 'AUTHOR COMMENT AND FUTURE PERSPECTIVE' SENTENCES**
>
> However, further research is required to ascertain if the environmental benefit associated with consumers being able to store and use more robust balloons for longer periods (waste reduction) is

offset by increased energy or chemical resource requirements during manufacturing or reduced biodegradability of the product at the end of its service life.

EXAMPLE CONCLUSION

Samples (e.g., balloons) that are typically *observation* (e.g., thermally and mechanically stable) for at least *period* (e.g., 24 months), are *observation* (e.g., more rapidly degraded) at *condition* (e.g., elevated temperatures) unless they are treated with *compound* (e.g., diimide). No *samples* (e.g., balloons) exhibited mechanical property degradation over the *test period* (e.g., 24 months) when stored at *condition* (e.g., ambient temperature) with *analysis* (e.g., scanning electron microscopy) indicating *observation* (e.g., no cracking) in these samples. Conversely, the *properties* (e.g., mechanical properties) of *samples* (e.g., balloons) exposed to *condition* (e.g., higher temperatures) rapidly deteriorated, with losses in *property* (e.g., tensile strength) and *observation* (e.g., cracking visible) during *analysis* (e.g., scanning electron microscopy). *Samples* (e.g., balloons) produced from *material* (e.g., latex) treated using *compound* (e.g., diimide) proved an exception to this rule, retaining their *properties* (e.g., tensile strength) even when stored at higher temperatures. The resilience of these *samples* (e.g., balloons) was attributed to *characteristic* (e.g., hydrogenation) resulting from *compound* (e.g., diimide) treatment, which *explanation* (e.g., increased the degree of saturation in the polymer and improved its resistance to heat). *Samples* (e.g., balloons) treated with *compound* (e.g., diimide) would *benefit and impact* (e.g., experience longer life spans, reducing waste and economic loss across an industry littered with bulk buying practices and discarded products no longer fit for purpose). However, further research is required to ascertain if the environmental benefit associated with consumers being able to store and use more robust balloons for longer periods (waste reduction) is offset by increased energy or chemical resource requirements during manufacturing or reduced biodegradability of the product at the end of its service life.

MATERIALS AND METHODS

The 'Materials and Methods' section of the manuscript, also known as materials and (experimental) methodology, experimental section or similar, describes in detail how experimental or analytical work was completed and with which resources. This section is very important to the scientific integrity of the manuscript. If the experimental work described in a manuscript cannot be reproduced by the scientific community with the same results, then its validity may be called into question. Care should subsequently be taken to ensure that all information provided in this section is accurate, detailed and fully transparent.

Materials and methods themed content can be likened to recipes in a cookbook with a writing style that sets it apart from the content appearing in other manuscript sections. Here, sentences take a highly instructional tone, are littered with details and smooth flow between sentences and topics is less important. Paragraphs are certainly themed but, like a cookbook, the stress is on simply and accurately describing what was done and how in a manner that enables someone else to reproduce the experimental or analytical procedures described in the manuscript in exactly the same way that they were performed by the authors.

Allocating Content to the Materials and Methods

In some cases, the line between 'Materials and Methods' and 'Results and Discussion' can be blurred, but in general the idea is to present the experimental or analytical results in a direct, matter of fact, fashion and discuss them without veering off course. Any content that relates to 'how' or 'why' an experiment was performed, something was selected or similar, is often best suited to the 'Materials and Methods' section of the manuscript to help avoid these detours from efficient topic flow in the 'Results and Discussion' section.

The bulk of all equations in a scientific manuscript also typically appear in the 'Materials and Methods' section. This is because equations are often used in conjunction with instrument output data to obtain useful new information on relationships, related parameters etc. Their use subsequently often constitutes a data processing and analysis methodology. It is rare that equations belong in another section of the manuscript unless the equation itself is the focus of the discussion.

In terms of writing order, there is no benefit in completing the 'Materials and Methods' before or after any other manuscript section and it can hence be written at any stage in the writing process provided all experimental and analytical work has been completed. Some people like to write it fairly early on to get it out of the way but, given its lack of relevance to the workflow suggested in this guide, it makes sense to approach it towards the end of the writing process. Provided all experimental work is well documented in a lab journal and will subsequently not be forgotten, there is no disadvantage in this strategy.

The 'Materials and Methods' section of a manuscript is typically broken into a 'Materials' subsection followed by themed sample preparation and experimental or analytical subsections each with a unique heading. These headings should use full sentences that provide all necessary experimental context e.g., 'morphology of the balloons' rather than simply 'morphology'. All relevant experimental or analytical information relating to the subsection theme can then be arranged below each heading in paragraphs.

Materials

In many scientific disciplines the first subsection to appear in the 'Materials and Methods' is entitled 'Materials'. This subsection constitutes little more than a list of all materials that were used at any point in the experimental work, including organisms, chemicals, samples, anything used to produce or process samples, instrument test standards or reference materials.

Items purchased from or donated by companies should be accompanied by the company name and location, often in brackets. It is also worth noting whether an item was 'purchased from' or '(kindly) provided or supplied by' a company. Relevant product information such as catalogue numbers, trade names, chemical grades (e.g., analytical grade), species and strain names should also be provided.

Materials collected from nature should be accompanied by a description of their source. Include all relevant information in this description, such as the collection site location, geographic coordinates, the collection date and season.

Review the manuscript and lab journal documenting all practical work completed over the course of the study and make a list of anything that was used that cannot be considered equipment or an instrument. Formulate sentences based on this list, starting with prefabricated samples or raw materials required to make samples, followed by anything needed to process them and finally anything used during testing or analysis including instrument test standards and references. Some indication should be provided as to which part of the experimental work each material belongs. If multiple materials within the same experimental phase come from the same place, then group them by the source.

EXAMPLE - MATERIALS REQUIRED IN BALLOON MANUFACTURING

Natural latex (*types*) was purchased from *company* (*city, country*) as the base material for balloon production. Zinc diethyldithiocarbamate (ZDC) accelerator was kindly supplied as an emulsion by *company* (*city, country*). Zinc oxide (ZnO) activator and crushed mica colouring agent were purchased from *company* (*city, country*). Calcium nitrate tetrahydrate and ethanol were purchased for use as coagulant from *company* (*city, country*). Talcum powder was purchased for use as a dry lubricant from *company* (*city, country*). All materials were used as received. Deionised water was used for all experiments.

Sample Preparation

Most scientific manuscripts feature some degree of sample preparation, be it composite fabrication in materials science or engineering, synthesis in chemistry or culturing or germinating seeds in biology. Any work required to produce a sample that will be tested or analysed in the manuscript must be thoroughly documented in a relevant subsection or subsections of the 'Materials and Methods'.

Review the manuscript and lab journal documenting the experimental work and formulate simple blunt sentences that comprehensively describe the exact processes used to produce the samples, step by step, detail by detail. Consider every sentence, the instructions and information it provides carefully. Could you produce identical samples using nothing but your own instructions? It can help to have someone unfamiliar with the process read your materials and methods and comment on anything that is unclear to them since they can more readily identify missing details or unclear information otherwise considered assumed knowledge by the author.

Information that could appear in a sample preparation themed subsection of the 'Materials and Methods' includes:

- Material quantities (masses, volumes, dimensions) and tolerances.
- Material concentrations, consistencies and mixing ratios.
- The duration of processes and time elapsed before, after or between them.
- Equipment used during production or processing.
- Equipment properties e.g., pore size.
- Equipment settings e.g., speeds, frequencies, temperatures, pressures.
- Storage or growth conditions and duration.

Details on equipment used should generally be accompanied by model number, manufacturer and their location. This is, however, not necessary for standard laboratory equipment, such as glassware, ovens and hand tools. These can be referred to generically. Be sure to capitalise equipment that is named after someone correctly e.g., Petri dish, Büchner funnel, Fourier transform (infrared spectrometer).

Since all materials used in the sample preparation process have already been comprehensively described in the 'Materials' subsection, there is no need to provide further details on these materials. Simply ensure that it is clear which previously described material is being referred to through the use of consistent labelling.

The same is true of previously described sample preparation processes. If multiple sample preparation subsections exist and some contain repeated steps, provide a full description of the process upon its first appearance in the text and then simply refer back to this description upon each repetition. If a process similar or identical to that used in a previously published scientific document is adopted, the reader can instead be referred to the published method.

This can be achieved with a statement such as "*sample* was prepared or processed as previously described for *sample*" or "in *reference*". If the process has been modified in some way, this should be noted and, if there is potential for confusion, then a brief summary can be included to provide clarification and context for the statement. This can be signposted with the term "briefly" followed by a short explanation of the modified process.

EXAMPLE - BALLOON PREPARATION PROCESS

A *mass* of natural latex was mixed with *mass* of ZDC accelerator, *mass* of ZnO activator and *mass* of crushed mica colouring agent to prepare a dipping solution. This was then filtered through a cheese cloth with a *pore size* to remove lumps and agitated for *period* (e.g., 15 h) to evenly distribute the colour and prevent the latex from congealing. Balloon forms were cleaned though immersion in a hot water bath at *temperature*. The clean forms were then dipped into a coagulant solution of calcium nitrate tetrahydrate dissolved in ethanol (e.g., 35% w/v) to coat them and give them an electrochemical charge to improve adhesion between the latex and the form before being dipped into the latex solution. The latex coated forms were again submerged in coagulant to remove any excess *compound* and ensure an even surface finish. The latex was allowed to air dry for *time* at *temperature* (e.g., 20 °C) and *humidity* (e.g., 30% RH). Lips were then manually rolled onto the balloons, and they were soaked in a water bath at *temperature* for *period* (e.g.,

16 mins) to remove impurities and vulcanise the latex material. The balloons were finally dipped in a dispersion of talcum powder (*percentage* w/v) and blown from their forms using compressed air. Following production, the balloons were again immersed in *temperature* (e.g., 70 °C) water to clean them and complete vulcanisation.

Experimental and Analytical Methods

All original research manuscripts contain either experimental work or analytical work or both. This could be as simple as physical experiments with hand recorded measurements like chemical titrations or biological growth rate measurements. Or it could be as complicated as instrument-aided characterisations using microscopes, spectrometers, chromatographs and other testing apparatus or even simulations and modelling that are entirely computer-based.

Experimental and analytical methods are best described in groups using themed subsections e.g., morphology, chemical composition, molecular structure, physical and mechanical properties of the balloons with all relevant methods described in paragraphs beneath the relevant group heading. This reduces the number of subsections required in the 'Materials and Methods' section, keeps all related methods together and correlates better with the way in which content is presented in the 'Results and Discussion' section than if every single method and instrument were to be allocated its own eponymously titled section.

As already noted, if any experimental or analytical process has already been described or published then it can be referred to with relevant modifications stated rather than repeated in full.

These themed experimental and analytical methods sections could include the following information:

- Instrument names, models, manufacturers and their locations (also for attachments).
- Any process that was required to prepare samples for testing.

- Sample masses, volumes, dimensions or other properties where relevant to the instrument, testing or analysis.
- The number of replicate tests or replicate samples tested.
- Test duration where relevant.
- Laboratory environmental conditions (e.g., temperature, humidity) for sensitive instruments.
- Instrument parameters, settings, ranges, limits and conditions.
- Instrument reference standards or reference materials.
- International experimental standards with which the testing complies (e.g., ISO, ASTM).
- Models, equations, algorithms, statistics, or computer software used during analysis.

EXAMPLE - EXPERIMENTAL AND ANALYTICAL METHODS FOR BALLOON TESTING

Morphological and elemental analysis of the balloons

Scanning electron microscopy (SEM) imaging and energy dispersive spectroscopy (EDS) elemental analysis of each balloon were performed using a *brand and model* scanning electron microscope with a *brand and model* energy dispersive X-ray spectrometer attached. An accelerating voltage of *value* (e.g., 30 kV) was used. Samples were air dried, and chromium coated (*model, manufacturer, city, country*) for *period* (e.g., 30 s) at *current* (e.g., 80 mA) prior to imaging. The EDS spectra of uncoated samples were also analysed for inorganic residues using the *brand and name* software (*version number*). An average spectrum was obtained based on individual spectra from *value* (e.g., 12) different sites. C, H, N, S and O elemental analysis was completed to supplement the EDS spectra for triplicate *mass* (e.g., 2 mg) samples using a *brand and model* CHNS–O elemental analyser.

Analysis of the molecular structure of the balloons

IR spectra were recorded using a *brand and model* Fourier transform infrared (FT-IR) spectrometer with a single reflection diamond attenuated total reflectance (ATR) module and KBr optics (beam splitter). Three spectra were recorded from different portions of material to verify homogeneity. Spectra were recorded across the full accessible range (e.g., from 4,000-400 cm^{-1}).

PHYSICAL AND MECHANICAL ANALYSIS OF THE BALLOONS

The balloons were cut into test specimens (shape according to *specimen type, standard*) using a *brand and model* manual cutting press. Specimens had *dimensions*. The thickness of each specimen was determined using a *brand and model* digital outside micrometer. Tensile tests were performed using a *brand and model* universal testing system, equipped with an e.g., 1 kN load cell and a *brand and model* non-contact video extensometer. Specimens were fixed between metal clamps using blotting paper to avoid perforating them. Testing velocity was *velocity* with gauge length set to *length*. The elastic modulus (E) was analysed in the linear elastic region as a secant between strength values separated by *percentage* strain. The tensile strength (σ_{UTS}) was calculated from maximum load and specimen cross-sectional area. Average values and associated standard deviation were calculated based on *value* (e.g., 7) replicates of the same sample type.

THERMAL DEGRADATION ANALYSIS OF THE BALLOONS

Balloon thermal degradation properties were assessed using a *brand and model* thermogravimetric analyser. Balloon fragment samples of ~*mass* (e.g., 10 mg) were placed in a platinum high temperature crucible and heated from *temperature* (e.g., 30 °C) to *temperature* (e.g., 600 °C) at a heating rate of *rate* (e.g., 20 °C/min). Samples were tested in air and nitrogen atmospheres (both 25 mL/min).

INTRODUCTION

Introductions do not need to be excessively long or general in their scope. Their purpose is to concisely introduce the topic and its relevance, the current state of the literature on the topic, the motivation for the study, i.e., a problem or research gap that needs to be solved or filled and brief details on the intended execution i.e., aims, materials and methods. However, the most efficient writing order for this content is not the order in which it appears. The following structured process can be used to confidently write a concise, precise and carefully scoped introduction for a bachelor's thesis, a master's or doctoral thesis research chapter, a journal research article or conference paper. Much of the structure also holds true for journal review article or book chapter introductions (see LITERATURE REVIEW MANUSCRIPTS for more details on differences). The introductory chapter of a doctoral thesis is formatted a little differently and is discussed in MONOGRAPH STYLE BACHELOR'S, MASTER'S OR DOCTORAL THESIS — D. THESIS INTRODUCTION CHAPTER.

Final Paragraph - Aims and Execution

Having already written the 'Results and Discussion', 'Conclusion' and 'Materials and Methods' sections of the manuscript (or chapter), the easiest starting point for the 'Introduction' is the aims and execution paragraph. One of the main reasons to write the introduction after the main body of the manuscript is already complete is that its content must closely match that of the experimental work, results, and discussion. Writing the introduction before the main body of the manuscript is finished can easily lead to discrepancies between the aims and experimental execution described in the introduction and what is actually presented in the results and discussion later in the manuscript text. However, with the results and discussion already written it is now very easy to read through the draft and decide what the study actually achieved and to derive appropriate aims from this. It should be noted the recommendation here is not to complete scientific work without aims and then to decide on the aims later, but rather that experiments can have unpredictable outcomes and that the aims originally conceived during the design of experiment phase might not necessarily match the final results and discussion obtained from the study, subsequently requiring modification.

Once the successfully demonstrated aims have been noted, continue on within the same paragraph to describe the execution of the study i.e., materials and methods. The already completed 'Materials and Methods' section of the manuscript can make describing the execution of the experimental work much easier. However, be careful not to get bogged down on details. The description of how the study was executed should be very concise, i.e., two or three sentences, and should maintain broad perspective, summarising the experiment very simply. Use the headings in the 'Materials and Methods' section of the manuscript in conjunction with the figures and tables in the 'Results and Discussion' section as a guide. Decide on the most important samples, variables, processes, instruments, methods, and analysis used to generate these results and be sure to include them in your description. Use broad categorical descriptions, such as morphology, composition, mechanical properties, structure rather than detailed information or examples. These sentences conclude the final paragraph of the introduction.

EXAMPLE - INTRODUCTION 'AIMS AND EXECUTION' PARAGRAPH

This study investigated the effect of *variables* (e.g., aging on the shelf life of untreated and diimide treated (hydrogenated) natural rubber balloons stored for long periods under ambient and elevated temperature conditions). *Samples* e.g., (balloons) were *methods* (e.g., manufactured and aged over a period of 24 months at various temperatures). The *measured data and analysis* (e.g., morphology and chemical composition of the balloon material were then assessed for all samples, and their mechanical properties evaluated).

Preceding Paragraphs

The paragraphs preceding the final paragraph must now be written. These paragraphs will introduce the topic of the manuscript and its relevance, the current state of the literature on the topic and the motivation for the study, i.e., a problem or research gap that needs to be solved or filled. This can be as long or as short as necessary but can generally be achieved in two to three substantial paragraphs. The scope of the content should be restricted to information that strictly relates to the aims, materials and methods detailed in the final paragraph of the introduction, and by association the manuscript, rather than general content on the research area. It is for this reason that the final paragraph of the introduction was written before these paragraphs.

Topic and Relevance

Introduce the topic based on the aims described in the final paragraph of the introduction. What research area do these aims fall within? What basic information does one need to know about this research area to understand how the current study fits into it and why it is valuable? Provide as much background information as necessary but remain concise and on topic. Background information could include the relevance of structures, organisms, reactions, materials, processes, applications, specific parameters from the materials and methods to the study.

It is, however, important to remember that the background information is aimed at the inexperienced reader so at the very least ensure that this part of the manuscript is accessible to everyone. Prioritise the expression of ideas in general terms with simple examples that can be understood without topic-specific knowledge and avoid technical or themed language or examples. Also, be careful not to include irrelevant information that will misdirect the reader or compromise the clarity of the manuscript focus.

If the aim of the research is to investigate the effect of aging on the shelf life of untreated and diimide treated (hydrogenated) natural rubber balloons stored for long periods under ambient and elevated temperature conditions, it is not necessary to provide a summary of the history of balloons. It is not important to describe the available colours or price differences between retailers. Avoid unnecessary information.

Some information has the potential to be useful but can risk pulling you off topic. For example, it is not important to discuss parties (application) or balloon production volumes (demand for the product from consumers) unless it is to demonstrate the impact and value of the research. It is not important to describe balloon manufacturing or shape unless these factors may affect aging behaviour and balloon integrity e.g., manufacturing faults causing deterioration of mechanical properties over time, various shapes being more susceptible to stress concentrations than others. Be careful with such topics. Ensure that you only discuss them if they are relevant to the aims and content of the manuscript and, even if they are, only touch on the most highly relevant aspects of these topics.

The critical background information relating to the effect of aging on the shelf life of untreated and diimide treated (hydrogenated) natural rubber balloons stored for long periods under ambient and elevated temperature conditions is really:

- The material from which they are produced.
- How this material ages.
- The effect of temperature on that process.
- Relevant changes to the material and process resulting from diimide treatment.
- Typical storage conditions for balloons.
- And perhaps the minimum acceptable volume and pressure of air required in a balloon for use and the mechanical properties associated with this condition.

These should be the central points of a concise and focussed introduction on this topic.

EXAMPLE - INTRODUCTION 'TOPIC AND RELEVANCE' PARAGRAPH

Balloons are a common decoration at parties and festivities across the globe [*reference*]. Made from *materials* (e.g., natural rubbers synthesised by members of the *Euphorbiaceae* or *Compositaceae* plant families), they can be filled with air to stretch to *value* times their original size [*reference*]. Most natural rubbers comprise *chemical structure* (e.g., isoprene units, each of which contains a single cis-configured double bond), and are *properties* (e.g., elastic with high mechanical strength) [*reference*]. However, natural rubbers are vulnerable to *stimuli* (e.g., heat, weather and chemicals) and their elasticity degrades with time [*reference*]. This degradation is the result of *chemical process* (e.g., generation of free radicals), which causes *explanation* (e.g., formation of new cross-links and oxidative hardening) [*reference*]. The stability of natural rubbers against *degradation processes* (e.g., thermal and oxidative degradation) can be improved by *chemical modification* (e.g., increasing the degree of saturation in the polymer (hydrogenation) using diimide, which provides hydrogen atoms to the C=C bonds) [*reference*]. This process provides resistance to elevated temperatures and electromagnetic radiation (e.g., light) that accelerate the aging process and could prevent improperly stored balloons (e.g., in direct sunlight) from rapidly degrading [*reference*].

Motivation and Research Gap

Now focus on the motivation and scope of the work. What problem are you trying to solve and how? Why are you and other researchers interested in solving this problem? What applications or implications does it have for society and the progression of science? What are the limits of what you can do? A few sentences are sufficient to provide the motivation, and scope of the research.

Support this with an overview of the current state of the literature. What have other researchers done to solve the problem so far? What has been successful? What are the limitations of the success? What has not been successful? What has not yet been attempted? Define the research gap and why it is important. At the manuscript preparation stage, the research gap should be very clear since it should be the subject of the experimental work and associated results and discussion addressing the successfully completed aims of the manuscript.

The discussion of the state of the literature should focus on publications that are as closely related to the defined research gap as possible. Try to focus on studies that directly border the hole that you intend to fill. How close this can be will depend on the research area and gap. The discussion of the state of the literature in sparsely populated research areas or for manuscripts that investigate truly unexplored ideas will obviously be a lot broader than that of a manuscript that focuses on tweaking or optimising parameters or ideas in established and popular research fields.

EXAMPLE - INTRODUCTION 'MOTIVATION AND RESEARCH GAP' PARAGRAPH

Although balloons are only expected to have a short service life once inflated, many consumers buy more balloons than they require at any one time and will store them for extended periods with the intention of using them again in the future [*reference*]. The compromised structural integrity of balloons stored for extended periods or under improper conditions results in many being thrown away [*reference*]. This constitutes a loss to the consumer and contributes to global waste management problems. Recent studies have investigated the use of *variations in manufacturing processes* to decrease stresses and thermal shock to balloons during production, thereby reducing breakage and losses [*reference*]. *Compounds and additives* have also been shown to improve the mechanical properties of natural rubber used in tyre manufacturing [*reference*] and hydrogenation of natural rubber using diimide is a well-established and optimised process [*reference*]. Modelling has even indicated that changes in the rate of inflation could help to make aged balloons useful again [*reference*]. However, very little attention has been afforded to examining the aging behaviour of balloons and what can be done to extend balloon service life under elevated temperature conditions.

EXAMPLE INTRODUCTION

When compiled these sections form the introduction. Always check the collated text thoroughly to ensure that it flows smoothly from topic to topic. Minor editing may be required to improve the narrative and make the transitions between the necessary points less abrasive to the reader.

Balloons are a common decoration at parties and festivities across the globe [*reference*]. Made from *materials* (e.g., natural rubbers synthesised by members of the *Euphorbiaceae* or *Compositaceae* plant families), they can be filled with air to stretch to *value* times their original size [*reference*]. Most natural rubbers comprise *chemical structure* (e.g., isoprene units, each of which contains a single cis-configured double bond), and are *properties* (e.g., elastic with high mechanical strength) [*reference*]. However, natural rubbers are vulnerable to *stimuli* (e.g., heat, weather and chemicals) and their elasticity degrades with time [*reference*]. This degradation is the result of *chemical process* (e.g., generation of free radicals), which causes *explanation* (e.g., formation of new cross-links and oxidative hardening) [*reference*]. The stability of natural rubbers against *degradation processes* (e.g., thermal and oxidative degradation) can be improved by *chemical modification* (e.g., increasing the degree of saturation in the polymer (hydrogenation) using diimide, which provides hydrogen atoms to the C=C bonds) [*reference*]. This process provides resistance to elevated temperatures and electromagnetic radiation (e.g., light) that accelerate the aging process and could prevent improperly stored balloons (e.g., in direct sunlight) from rapidly degrading [*reference*].

Although balloons are only expected to have a short service life once inflated, many consumers buy more balloons than they require at any one time and will store them for extended periods with the intention of using them again in the future [*reference*]. The compromised structural integrity of balloons stored for extended periods or under improper conditions results in many being thrown away [*reference*]. This constitutes a loss to the consumer and contributes to global waste management problems. Recent studies have investigated the use of *variations in manufac-*

turing processes to decrease stresses and thermal shock to balloons during production, thereby reducing breakage and losses [*reference*]. *Compounds and additives* have also been shown to improve the mechanical properties of natural rubber used in tyre manufacturing [*reference*] and hydrogenation of natural rubber using diimide is a well-established and optimised process [*reference*]. Modelling has even indicated that changes in the rate of inflation could help to make aged balloons useful again [*reference*]. However, very little attention has been afforded to examining the aging behaviour of balloons and what can be done to extend balloon service life under elevated temperature conditions.

This study investigated the effect of *variables* (e.g., aging on the shelf life of untreated and diimide treated (hydrogenated) natural rubber balloons stored for long periods under ambient and elevated temperature conditions). *Samples* e.g., (balloons) were *methods* (e.g., manufactured and aged over a period of 24 months at various temperatures). The *measured data and analysis* (e.g., morphology and chemical composition of the balloon material were then assessed for all samples, and their mechanical properties evaluated.)

Selecting References for the Introduction

Selection of optimal references is very important in the introduction. This is because, as the section that opens the manuscript, it will be the first impression that a reader, including an examiner, reviewer, or editor, has of the manuscript. Depending on the referencing style used in the manuscript, the authors and publication years of the references that are being cited may be visible within the highly conspicuous introduction text. Even if they are not, references appearing in the introduction can still draw attention through first appearance at the top of the manuscript reference list if a referencing style that uses sequential numbering is adopted.

It is obviously important to always use the most relevant and appropriate references in the manuscript. Reference selection should never be guided by anything other than the quality of the scientific work being cited and its suitability in supporting the discussion that is being developed. However, it is worth at least considering the implication behind the use of certain references. For example, the use of many very recent references in the introduction demonstrates the current timeliness of the research. It shows that there are many other people working on the same topic and that it is subsequently currently a topic of value and interest to the scientific community.

If appropriate, citing articles from high impact journals or prominent authors in the field in the introduction can be used to similar effect. The ability to do so can help to demonstrate that the topic is of interest to highly selective journals and esteemed researchers and lends credibility to the concept that the research is important and valuable. When writing journal articles, if the target submission journal has previously published content related to the research, citing some of their relevant and appropriate publications can also help to demonstrate that your manuscript falls within their scope.

ABSTRACT

The most important attribute of a scientific document is the new knowledge that it details, which represents its scientific value and contribution. But unless this value can be clearly and efficiently conveyed to the target audience, the impact of even a ground-breaking discovery can be severely diminished. The volume of documents being published every year coupled with those already in existence can make attracting the right readership quite challenging. For this reason, having a well written abstract that very precisely reflects the core aspects of the document is very important.

An abstract is essentially a very concise summary of each section of the scientific manuscript and is often the only part of the document that a potential reader will check before deciding whether it is of interest and use to them or not. It is also often the only part of an article that is free to view for those who do not have subscription access to journal content. Researchers searching through hundreds of documents for information that they need will likely only afford seconds to glance at the abstract before moving to the next. An abstract subsequently needs to be the living embodiment of the document in a tiny fraction of the space. Word limits for abstracts vary but they are generally very short i.e., up to a few hundred words. For journals, the abstract word limit is often strictly enforced and should be checked in the journal's Guide for Authors (see PUBLISHING JOURNAL ARTICLES — FORMATTING A MANUSCRIPT FOR SUBMISSION — USING THE GUIDE FOR AUTHORS). Thesis abstracts are usually more flexible and can consist of a main abstract for the thesis but potentially also individual abstracts for each chapter.

A thesis abstract is usually mandatory and is discussed in the USING THIS GUIDE – MONOGRAPH STYLE BACHELOR'S, MASTER'S OR DOCTORAL THESIS — E. THESIS ABSTRACT section of this guide. Writing an abstract for each research chapter is optional but can help to improve the accessibility of information within the thesis and serve as inspiration for text in the thesis abstract and the 'thesis outline' for the thesis 'Introduction' chapter. Even if you decide not to include chapter abstracts in the final version of your thesis, it is worth writing and holding onto them to help with these processes later. The structured thesis writing process described in this guide assumes that an abstract will be produced for each research chapter in addition to a main abstract for the thesis. This section focuses on writing an abstract for an original research thesis chapter, journal article or conference paper. Note: Inclusion of references in the abstract should be avoided.

The appropriate content for an abstract and its distribution could resemble the following:

- Topic and relevance (background information and context) (1-2 sentences).
- Motivation and research gap (1 sentence).
- Aims and execution (materials and methods) (1 sentence).
- Key results (3-5 sentences).
- Manuscript value and impact (1-2 sentences).

To prepare an abstract, start by pasting the manuscript's 'Introduction' and 'Conclusion' into a new Microsoft Word document. If written correctly, these sections already contain most of the content required for the abstract. The abstract's 'topic and relevance', 'motivation and research gap' and 'aims and execution' sentences come from the 'Introduction'. The 'key results' and 'manuscript value and impact' sentences come from the manuscript's 'Conclusion' supplemented by the 'Results and Discussion' section where necessary.

While it may be tempting to simply edit the pasted material as necessary to form an abstract, this is often an ill-advised approach. Most people grow attached to their own writing and their wording for specific sentences. It can subsequently be challenging to cut the contents of the introduction and conclusion to the word count required for the abstract and to make the abstract text sound considerably different to these other sections. An abstract that too closely resembles the introduction or conclusion not only sounds repetitive but no longer fulfills its freestanding image and function.

A better approach is to use the introduction and conclusion as inspiration for new text. Select the abstract sentence that will be written and use the manuscript's introduction or conclusion to identify appropriate content to satisfy the requirements of that sentence. Then return to the draft abstract and write the new sentence from memory based on what was just read. Since an abstract is generated based entirely on previously written material, there is little advantage in writing it in any order other than the order in which the sentences appear.

Topic and Relevance

Background information must be provided to introduce the topic of the manuscript. However, this information should be as brief and concise as possible. One potentially critical mistake made when writing this part of the abstract, as is also often the case in introductions, is that these sentences do not align closely enough with the focus of the study. They are too vague, focus on the wrong aspect of the study or potentially even completely the wrong topic.

Even when using a well-scoped introduction as inspiration, it is important to formulate the introductory sentences of the abstract carefully to provide highly precise context and background information on the study. While more leeway is possible in the introduction, an abstract has such a restrictive word limit that these sentences must be perfectly formulated to express only the most critical and relevant details. They should be carefully focussed on the central theme of the manuscript.

Carefully consider the background information in the first paragraph of the manuscript's introduction. Which points are most critical to understanding the topic and its relevance? Which are less important? If the topic of the manuscript is evaluating the effect of aging at various temperatures on the material properties of untreated and diimide treated (hydrogenated) natural rubber balloons, it is important to introduce the balloon material and its value to balloon function in the first sentence.

This is then followed by a sentence linking the background information to the problem definition, which describes the relevance of the study. If there is no problem to be solved, then the study is not worth doing. A clear correlation should be visible between the first sentence containing background information (topic) and the second sentence describing the problem (relevance). If the flow from one sentence theme to the other is not apparent, then one of the sentences is not formulated correctly. The incorrect sentence is more likely to be the background information sentence since wandering off topic is quite easy to do, while misidentifying the problem is rarer.

EXAMPLE - ABSTRACT 'TOPIC AND RELEVANCE' SENTENCES

Introduction – 'topic and relevance' paragraph:

Balloons are a common decoration at parties and festivities across the globe [*reference*]. Made from *materials* (e.g., natural rubbers synthesised by members of the *Euphorbiaceae* or *Compositaceae* plant families), they can be filled with air to stretch to *value* times their original size [*reference*]. Most natural rubbers comprise *chemical structure* (e.g., isoprene units, each of which contains a single cis-configured double bond), and are *properties* (e.g., elastic with high mechanical strength) [*reference*]. However, natural rubbers are vulnerable to *stimuli* (e.g., heat, weather and chemicals) and their elasticity degrades with time [*reference*]. This degradation is the result of *chemical process* (e.g., generation of free radicals), which causes *explanation* (e.g., formation of new cross-links and oxidative hardening) [*reference*]. The stability of natural rubbers against *degradation processes* (e.g., thermal and oxidative degradation) can be improved by *chemical modification* (e.g., increasing the degree of saturation in the polymer (hydrogenation) using diimide, which provides hydrogen atoms to the C=C bonds) [*reference*]. This process provides resistance to elevated temperatures and electromagnetic radiation (e.g., light) that accelerate the aging process and could prevent improperly stored balloons (e.g., in direct sunlight) from rapidly degrading [*reference*].

Example abstract 'topic and relevance' sentences:

Balloons are produced from highly elastic *materials* (e.g., latex) and derive their aesthetic value from their ability to be inflated. However, *condition* (e.g., elevated temperatures) can cause *chemical process* (e.g., oxidative hardening) in unsaturated natural rubber polymers, compromising balloon integrity and subsequently their value.

Motivation and Research Gap

Although the second paragraph of the manuscript introduction deals quite broadly with the problem definition and current state of the literature, the 'research gap' sentence in the abstract must be very concise and to the point. Forget the extended discussion and details on what other researchers have done to date. What isn't known? Focus on the research gap and describe it as simply and efficiently as possible. The language should be accurate and specific but since the next sentence in the abstract will describe the 'execution of the study' in some detail there is no need to waste highly limited words at this point trying to comprehensively define potentially quite complex research gaps.

Where appropriate, it is also important to provide context within the sentence to illustrate the importance of the research gap and why it is worth investigating. If the research gap is that the aging behaviour of balloons and how it can be slowed are not well characterised, especially at elevated temperatures, then why should we bother to fill this gap? What is the impact of this problem? Linking the research gap that you are investigating to its relevance to science and society helps to demonstrate the value of the manuscript.

EXAMPLE - ABSTRACT 'MOTIVATION AND RESEARCH GAP' SENTENCE

Introduction – 'motivation and research gap' paragraph:

Although balloons are only expected to have a short service life once inflated, many consumers buy more balloons than they require at any one time and will store them for extended periods with the intention of using them again in the future [*reference*]. The compromised structural integrity of balloons stored for extended periods or under improper conditions results in many being thrown away [*reference*]. This constitutes a loss to the consumer and contributes to global waste management problems. Recent studies have investigated the use of *variations in manufacturing processes* to decrease stresses and thermal shock to balloons during production, thereby reducing breakage and losses [*reference*]. *Compounds and additives* have also been shown to improve the mechanical properties of natural rubber used in tyre manufacturing [*reference*] and hydrogenation of natural rubber using diimide is a well-established and optimised process [*reference*]. Modelling has even indicated that changes in the rate of inflation could help to make aged balloons useful again [*reference*]. However, very little attention has been afforded to examining the aging behaviour of balloons and what can be done to extend balloon service life under elevated temperature conditions.

Example abstract 'motivation and research gap' sentence:

Despite the fact that balloons are often purchased in bulk and stored for extended periods very little is known about their aging behaviour and how to slow it, especially at elevated temperatures.

Aims and Execution

The 'aims and execution' paragraph of the manuscript's introduction is already typically short and to the point, making the formulation of a representative sentence for this section of the abstract fairly straightforward. Nonetheless, the language used in this paragraph must be sharpened to accommodate the abstract's low word count.

Select words carefully to describe the execution of the study with maximum efficiency. Only include details that are critical to a fundamental understanding of the experimental work performed in the study. Use umbrella terms and simplification where necessary to convey the central themes of the experiment. Supplement this broader language with carefully selected technical terms and adjectives to hone the accuracy of the sentences to the experimental work while maintaining a sufficient level of accessibility for a broad audience e.g., mechanical properties instead of tensile tests, morphology instead of scanning electron microscopy.

EXAMPLE - ABSTRACT 'AIMS AND EXECUTION' SENTENCE

Introduction – 'aims and execution' paragraph:

This study investigated the effect of *variables* (e.g., aging on the shelf life of untreated and diimide treated (hydrogenated) natural rubber balloons stored for long periods under ambient and elevated temperature conditions.) *Samples* (e.g., balloons) were *methods* (e.g., manufactured and aged over a period of 24 months at various temperatures.) The *measured data and analysis* (e.g., morphology and chemical composition of the balloon material were then assessed for all samples, and their mechanical properties evaluated).

Example abstract 'aims and execution' sentence:

This study investigated changes in the *measured data and analysis* (e.g., morphological, chemical and mechanical properties) of *samples* (e.g., untreated and diimide treated (hydrogenated) natural rubber balloons) at *conditions* (e.g., a period of 24 months at various temperatures.)

Key Results

Despite representing just one of the five points covered in the abstract, the 'key results' sentences can constitute the bulk of its content, typically encompassing a third to as much as half of the abstract word count. Since the 'Results and Discussion' section of the manuscript represents the bulk of its volume and scientific value, this is hardly surprising. The other abstract sentences play supporting roles in introducing, summarising, and providing context for the 'key results' sentences in the same way that their respective sections do in the manuscript.

The 'key results' sentences of the manuscript's conclusion should already provide a concise and highly scoped overview of only the most important manuscript results and provide great inspiration for new sentences summarising the key results in the abstract. But again, even if these sentences were formulated in a fashion that was sufficiently concise for reuse in the abstract, which is unlikely, this should be avoided.

The manuscript's conclusion has already moved away from a results and discussion style interpretation, comparison and explanation of variables, samples and literature to focus on the big picture and use trends, correlations or relationships between the samples, properties, or parameters to define the main results which support the conclusion and 'central message' of the manuscript. Take it a step further now. Try to consolidate the key results into a few short, themed sentences which hammer home the 'central message' of the manuscript and are supported by the most important results from the manuscript's conclusion.

EXAMPLE - ABSTRACT 'KEY RESULTS' SENTENCES

Conclusion – 'key results' sentences:

No *samples* (e.g., balloons) exhibited mechanical property degradation over the *test period* (e.g., 24 months) when stored at *condition* (e.g., ambient temperature) with *analysis* (e.g., scanning electron microscopy) indicating *observation* (e.g., no cracking) in these samples. Conversely, the *properties* (e.g., mechanical properties) of *samples* (e.g., balloons) exposed to *condition* (e.g., higher temperatures) rapidly deteriorated, with losses in *property* (e.g., tensile strength) and *observation* (e.g., cracking visible) during *analysis* (e.g., scanning electron microscopy). *Samples* (e.g., balloons) produced from *material* (e.g., latex) treated using *compound* (e.g., diimide) proved an exception to this rule, retaining their *properties* (e.g., tensile strength) even when stored at higher temperatures. The resilience of these *samples* (e.g., balloons) was attributed to *characteristic* (e.g., hydrogenation) resulting from *compound* (e.g., diimide) treatment, which *explanation* (e.g., increased the degree of saturation in the polymer and improved its resistance to heat).

Example abstract 'key results' sentences:

Samples (e.g., balloons) exposed to *condition* (e.g., higher temperatures) *observation* (e.g., mechanically deteriorated), exhibiting reduced *property* (e.g., strength), and *observation* (e.g., visible cracking) during *analysis* (e.g., morphological analysis). Balloons produced from *material* (e.g., latex) treated with *compound* (e.g., diimide) were the only exception to this rule, likely attributable to their *property* (e.g., higher degree of hydrogenation).

Manuscript Value and Impact

The value and impact of a manuscript's scientific contribution should be briefly highlighted at the end of the abstract. The 'manuscript value and impact' sentences from the manuscript's conclusion offer great inspiration for this final sentence.

Rephrase the sentences to fit the abstract, ensuring that the most important, highest impact elements are strongly emphasised. Remember, the abstract is the only part of the manuscript that most people will read, and this sentence is the only opportunity to spell out why the manuscript is so important, so valuable, so worth reading.

EXAMPLE - ABSTRACT 'MANUSCRIPT VALUE AND IMPACT' SENTENCE

Conclusion – 'manuscript value and impact' sentences:

Samples (e.g., balloons) treated with *compound* (e.g., diimide) would *benefit and impact* (e.g., experience longer life spans, reducing waste and economic loss across an industry littered with bulk buying practices and discarded products no longer fit for purpose).

Example abstract 'manuscript value and impact' sentence:

Such compounds could considerably improve the life of the hundreds of millions of balloons sold each year, reducing the number of unused balloons discarded and yielding positive economic and environmental impact across the globe.

EXAMPLE ABSTRACT

Balloons are produced from highly elastic *materials* (e.g., latex) and derive their aesthetic value from their ability to be inflated. However, *condition* (e.g., elevated temperatures) can cause *chemical process* (e.g., oxidative hardening) in unsaturated natural rubber polymers, compromising balloon integrity and subsequently their value. Despite the fact that balloons are often purchased in bulk and stored for extended periods very little is known about their aging behaviour and how to slow it, especially at elevated temperatures. This study investigated changes in the *measured data and analysis* (e.g., morphological, chemical and mechanical properties) of *samples* (e.g., untreated and diimide treated (hydrogenated) natural rubber balloons) at *conditions* (e.g., a period of 24 months at various temperatures.) *Samples* (e.g., balloons) exposed to *condition* (e.g., higher temperatures) *observation* (e.g., mechanically deteriorated), exhibiting reduced *property* (e.g., strength) and *observation* (e.g., visible cracking) during *analysis* (e.g., morphological analysis). Balloons produced from *material* (e.g., latex) treated with *compound* (e.g., diimide) were the only exception to this rule, likely attributable to their *property* (e.g., higher degree of hydrogenation). Such compounds could considerably improve the life of the hundreds of millions of balloons sold each year, reducing the number of unused balloons discarded and yielding positive economic and environmental impact across the globe.

SELECTING A TITLE

Manuscript titles are the very first thing that readers see when searching through literature for material relevant to their research and are also important for indexing by search engines. The best titles are those which most accurately reflect the manuscript's content while simultaneously being simple enough to maximise accessibility to readers of all backgrounds. They should also stand out sufficiently against a sea of other search results to ensure that the reader actually notices them, reads them properly and, having done so, ultimately wants to proceed to read the manuscript's abstract.

Titles should ideally be short, avoid abbreviations and jargon yet contain enough detail to accurately convey the manuscript's topic. This can be a difficult balance to strike. Start by formulating the most detailed and accurate title possible with little emphasis on length and audience accessibility. The manuscript's 'Central Message' can be a good starting point for this.

EXAMPLE - CENTRAL MESSAGE

Samples (e.g., balloons) that are typically *observation* (e.g., thermally and mechanically stable) for at least *period* (e.g., 24 months), are *observation* (e.g., more rapidly degraded) at *condition* (e.g., elevated temperatures) unless they are treated with *compound* (e.g., diimide).

Focus on the length first by cutting out any irrelevant words and trying to express the 'central message' as efficiently as possible. Play with the use of adjectives, adverbs, and word order to reduce the number of words in the sentence. The title should be no longer than 12 words with longer titles less likely to engage the audience. On the other hand, titles that are too short, e.g., 2-3 words, are often too vague to accurately convey the topic of the manuscript.

EXAMPLE - DETAILED TITLE

Compound (e.g., diimide) improves *properties* (e.g., balloon thermal and mechanical stability) during *phenomenon* (e.g., thermal aging).

It is perfectly acceptable to use a well-formed detailed title, and, in some cases, it can even be advantageous to do so. The more detailed the title is, the more likely the manuscript is to reach its target audience. However, there are almost always readers who do not fall within the target audience yet could still find the manuscript interesting or useful. Science is very interdisciplinary and even manuscripts with very specific or obscure themes can end up being used to support or inspire other research areas.

Titles should ideally make relevant manuscript content recognisable to researchers in the target research area while simultaneously being intelligible to a broader audience. In addition to minimising abbreviations and jargon, titles should ideally allude to a manuscript's impact and demonstrate its value to readers unfamiliar with the research area. The aim here is to provoke interest in the topic and perhaps even inspiration for interdisciplinary applications of the research. To improve audience reach and demonstrate impact, swap any subject specific terminology for more general language which reflects the real-life applications of the research appearing in the manuscript.

EXAMPLE - IMPACT TITLE

Compound (e.g., diimide) improves *property* (e.g., balloon lifespan) at *condition* (e.g., elevated temperatures).

Finally, select a title type for the manuscript. The number and hierarchy of title types is a matter of debate and various classification structures exist, but they can be broadly categorised into:

- Declarative titles, which supply readers with the conclusions of the manuscript immediately – e.g., Diimide increases balloon material lifespan at elevated temperatures.
- Descriptive titles, which introduce the manuscript topic but do not betray its conclusions – e.g., Effect of diimide on balloon life at elevated temperatures.
- Interrogative titles, which introduce the topic in the form of a question to pique reader interest – e.g., Balloon integrity: Is diimide the key to increased material lifespan at elevated temperatures?

KEYWORDS

Manuscripts are generally indexed based on their title, abstract and a set of author-defined keywords. The number of keywords required varies by publisher, but most scientific manuscripts, including theses, journal articles, conference papers and book chapters will use 4-8 keywords to provide an overview of the manuscript content.

Keywords must sometimes be chosen from a set list while in other cases authors have more freedom of choice. It is important to be familiar with publisher expectations on keyword selection before brainstorming appropriate options. Journal keyword requirements are usually described in the Guide for Authors (see PUBLISHING JOURNAL ARTICLES — FORMATTING A MANUSCRIPT FOR SUBMISSION — USING THE GUIDE FOR AUTHORS).

Try to scope keywords very carefully to the central themes and results of the manuscript. If you could choose just a few words to sum up the most important aspects of the study, what would they be? The emphasis here is not the entire manuscript but rather its 'central message', associated themes and story.

Keywords could relate to topics such as:

- Sample types that hold a central importance to the investigation.
- Highly relevant variables, behaviour or phenomena that were tested or observed.
- Analytical methods, models, theories that strongly relate to the manuscript story.
- Especially important testing, experimental methods and instruments used in the characterisation of samples and subsequent analysis.
- Processing methods and conditions that hold critical emphasis to the study.

Keywords normally appear on a single line separated by semicolons (;) or as a list with each on a new line. Be sure to write out any abbreviations or acronyms in full so that each individual word is indexed rather than the abbreviated version or letters. A single keyword can also comprise of multiple words, the first of which is capitalised.

Single words are often too vague to facilitate accurate indexing e.g., strength. They are too similar to keywords, or their related derivatives, used by countless other manuscripts for close matches with search criteria to be possible. Combinations of multiple words can provide more specific and relevant keywords.

Once keywords have been selected, they can be tested using an indexing service such as Google Scholar. Enter the keywords and check that the search results returned are of a similar theme to that of the prospective manuscript. If incorrect or too loosely related results surface, then be sure to refine the keywords until they are more strongly correlated with the target theme.

EXAMPLE - KEYWORDS

Keywords: Natural rubber latex balloons; Diimide-induced hydrogenation; Thermal aging; Service life; Tensile properties; Microcracking.

REFERENCES

Scientific inquiry is often much too complicated and expansive for any single researcher or even group of researchers to ever hope to address over the course of their lifetime let alone in a single manuscript. Research instead encompasses contributions from many scientists over many generations. It is for this reason that the publishing standards and institutions that exist today operate as they do. Without a proper method for documenting scientific discoveries, many important findings would be lost and forgotten over time. Nonetheless, such large-scale documentation inevitably requires organisational structures and systems of its own.

Scientific integrity is underpinned by the way in which facts are reported. Any scientific claim must be supported by experimental and analytical work that is openly communicated. Academics must take responsibility for their claims, disclose how experimental work was completed so that others can reproduce it if they wish and be prepared to provide their data to others and defend it both during and long after initial publication. Every scientist takes responsibility for their contribution and hence their piece in the scientific puzzle.

At the same time, it would be very inefficient if researchers had to prove absolutely everything in each study, including things that other scientists have already proven. This is where references come into play. Instead of performing an experiment, a researcher can indicate that the experiment or analysis was already performed by someone else by using a reference. This allows for more rapid scientific progress and prevents unnecessary duplication in the literature since, although an experiment detailed in a scientific manuscript may be performed many times by many different people, only a single document on the discovery need exist.

A reference is inserted immediately after any claim, explanation or discussion that occurs in a scientific manuscript for which supporting data is not provided in that manuscript. A full list of references is also provided at the end of the document. If no experimental work has been performed to support a claim and no other published material that can be referenced exists in the literature, then the claim cannot be substantiated and should be omitted from original research manuscripts. Exceptions exist in scientific manuscripts with provisions for more hypothesis and author opinion such as Perspective articles.

Many referencing styles exist with some well-known examples including MLA (Modern Language Association), APA (American Psychological Association), Chicago and Harvard referencing. Preferred referencing style varies by research area, institution and department. Many journals also use referencing systems that are modified to their requirements. Subsequently, the intricacies of each style will not be discussed in this guide. Universities often provide templates and documents describing the use of their preferred referencing style for students completing a thesis and details on journal referencing expectations are usually provided in the journal's 'Guide for Authors' (see PUBLISHING JOURNAL ARTICLES — FORMATTING A MANUSCRIPT FOR SUBMISSION — USING THE GUIDE FOR AUTHORS).

When selecting documents to reference in a manuscript the emphasis should be on using reliable, peer reviewed scientific publications. Journal articles are arguably the best references due to the scrutiny that they receive during the publication process. Books, conference papers, patents, and reports from well-known organisations are all also reviewed to some extent. Books are typically edited, conference abstracts at least checked by the organising committee, patents examined during the application process and reports internally reviewed by their issuing organisation.

Web resources, be it webpages or documents hosted online by members of the public, do not generally make good references. Some journals allow weblinks to be quoted in text in brackets as a supplement to traditional references but referencing websites in the traditional style should be avoided where possible. This is because, while published documents such as journal articles, conference papers, books and reports from organisations are stable and reliable references managed by responsible institutions with set procedures for making changes, such as registered revisions and editions, independent web content can change at any time. Webpages can be removed, updated, or altered within seconds by their administrator. The highly dynamic nature of web content can very quickly compromise its scientific integrity.

Reference Management Software

Although it is of course possible to insert and format references manually, as was tradition prior to the widespread use of computers, software now exists which considerably simplifies the referencing process. Popular examples include EndNote (www.endnote.com) and Mendeley (www.mendeley.com) reference managers. These software packages typically consist of an independent program where references can be added, removed, updated, and archived and an add-on for standard word processing software, such as Microsoft Word, so that users can access and use their reference libraries while writing. Most universities provide students with access to EndNote for free and Mendeley is free by default. Operating instructions for these software packages are available from their respective manufacturers and are not discussed in this guide.

The key benefits of using a reference manager as opposed to referencing manually are that references can be added and removed very easily at any point in time without having to reformat the existing references. This is especially beneficial for some sequentially numbered referencing styles, where updating existing references following even fairly minor modifications to a given reference can be very time consuming and complicated.

Using Exportable Citations

Using a reference manager also means that the vast quantity of stored information in indexing services can be exploited to accelerate the referencing process. In most cases it will not be necessary to enter details such as document authors or title into the reference manager. Simply find the document that will be referenced online and look for its exportable citation. Most journals will provide the option to download citation information for every article on their website for free. Exportable citations are normally indicated with a quotation mark ("). Simply click the symbol and select the desired reference manager format from the list and a file containing all available citation information for the document will be downloaded. Open the file to add the reference information to your reference manager.

If exportable citations are not provided by the publisher, then use those provided by Google Scholar (www.scholar.google.com). Simply enter the document title or other identifying features and complete a search for it. Once the desired document has been identified in the search results list, look below the title, author list and abstract excerpt to the blue links at the base of every search result. The exportable citation for the document is provided there and is depicted by the same quotation mark symbol (“) used by publishers.

One thing to be careful of when using exportable citations is the information that they contain for the year of publication, volume, issue, and page numbers. Journal articles are often published online before they are assigned these details meaning that this information may be initially unavailable. It is also possible to cite documents which have not yet been published, although they must at least have been accepted for publication by the journal. When citing such documents, it may not be possible to provide a volume, issue or page number(s) and the digital object identifier of the article (DOI) can be used in their place. This string of characters accompanies the bibliographic citation of most scientific electronic documents and serves as both an identifier and active link to the electronic content on the web. If an article that has been accepted but not yet published is being referenced, then this should be indicated in brackets i.e., (accepted) or (in press).

The year of publication included in exportable citations may also change. An exportable citation that is generated for an article released near the end of the year will use the year that it appeared online as the year of publication. If the article is later assigned to an issue that will be released in the following year, then the year of publication listed in the original exportable citation will be wrong. This is unavoidable in many cases but reinforces the fact that the information appearing in the exportable citations of recently published articles should be checked again later if they are to remain in the reference manager and be reused in future manuscripts.

Using Reference Style Templates

References inserted using a reference manager are added as software-controlled field codes rather than as plain text. They are subsequently very quickly, easily, and automatically modifiable. As individual references are inserted within the text body of the manuscript, a complete reference list is automatically generated at the end of the document and updated with each new addition. The referencing style that the reference manager is set to during the writing process is of little importance since it can easily be changed without affecting the other document content. This ability to easily switch referencing styles can be another very advantageous feature of reference management software when it comes to writing journal articles. Most theses use fairly standard referencing with the required style typically known prior to the thesis being written. Conversely, referencing styles used by journals can vary considerably and may differ from standard styles. This not only complicates initial manuscript preparation but can make changing referencing styles prior to resubmission to a second or third journal challenging if the manuscript is initially rejected.

These issues can be easily mitigated using reference management software. Reference managers, such as EndNote, are typically preconfigured with templates for many journals. Attaining the correct referencing style for an initial submission or changing style to suit a different journal during resubmission is as simple as selecting the journal from a list and waiting as the software modifies the in-text references and reference list at the end of the document accordingly. If a journal referencing style template is not included in the software by default, it can usually be downloaded from the website of the software manufacturer e.g., www.endnote.com/downloads/styles.

Converting to Plain Text

Once the manuscript text is finished, all references have been included and the correct referencing style selected, the final step is to convert the references in the document to plain text. As discussed, reference managers use field codes within word processing software documents to generate and change references. It is for this reason that referencing styles can be so easily switched and references updated or modified. However, as useful as this code is during the writing process, it must be converted to text that is not computationally tagged or specially formatted prior to submission to a third party.

Most reference management software makes this process very easy with options within the word processor add-on to convert field codes to plain text. It is however important to ensure that this occurs once the document is already finished and is the absolute last step completed prior to submission. The reason for this is that once the document has been converted to plain text the option to use the reference manager to quickly and easily update or modify references or referencing style will no longer be available. This can complicate revisions that involve reference modification, which are especially relevant during thesis examinations or the journal peer review process and when changing referencing style for resubmission to a different journal if the first submission is unsuccessful.

For this reason, it is very important to save a copy of the final manuscript with the reference manager code inside it directly before creating a plain text version of the manuscript for submission. This document can then be used for any revisions that need to be made in the future and a new plain text document produced from this master document for resubmission.

LITERATURE REVIEW MANUSCRIPTS

A literature review provides an extensive synthesis of an established research field, summarising the results and developments of many existing original research manuscripts in a single document. It acts as an independent introductory reference for those unfamiliar with a research area and is subsequently an important element in most theses, where a reader must first acquaint themselves with a research field before being able to understand and appreciate the value of a new contribution.

Literature reviews are also published periodically by journals to provide a moving synthesis and consolidation of research activity, and it is possible for the keen student to publish their literature review if they wish. If students intend to publish their literature review, it is important to remember that many review articles probably already exist on the same or similar topics and that, even if they don't, competition to publish is fierce. The ideas in the following section can help to improve any literature review, including those intended for a thesis or book chapter, but are most applicable to those wanting to publish a review article in a journal.

IMPROVING REVIEW ARTICLE IMPACT IN JOURNALS

Most journals accept review article submissions; however, some especially reputable journals will require authors to hold an invitation to do so. This requirement stems from journals wanting to steer the topic, quantity, and publication frequency of their reviews to match journal and research field requirements. They also prioritise reviews from esteemed academics and experts in the field, who have considerable insight and expertise to offer and are subsequently often invited to submit such articles. This invitation requirement should, however, not dissuade the skilled student from seeking to submit their well-researched and well written review to a journal.

Attaining an invitation to submit a review article is as simple as formulating and submitting a letter that convincingly pitches the review idea to a journal editor. A proposed abstract and list of recent related publications demonstrating the prospective author's prowess in the field may also be required. Students who do not yet have related publications may elect to submit their review to less prestigious journals that do not have an invitation requirement or team up with a more established academic who can solicit an invitation from an editor.

That said, the success of a review often hangs on the topic and the degree of novelty rather than the author's reputation. These in turn largely depend on how the author approaches the review. Just because a review utilises existing literature rather than presenting new experimental work does not mean that it should not have novel value, make a new contribution to science or be creative. Journals are overloaded with review manuscript submissions that are little more than lists recounting the contents of other research articles. Some also use so few references that the difference in time and efficiency required to read the abstracts of the articles included in the review and the review itself is not apparent. While some such review articles might make it through the peer review process depending on their topic and content, many will be rejected.

When writing a review article for a journal it is critical to always consider what the novel value and contribution of the review will be. The best place to start is to have a unique scope if possible. Try to approach the research field from a new angle. If a review has notably different content or focus to the many other reviews in the area, then it stands a much better chance of being accepted.

It is also critical to add value to the literature that is included in the review. Do not simply recount what other people did. If a review starts to move from study to study, then this should immediately ring alarm bells. Every section should instead focus on contrasting a broad range of literature with a central theme before moving to the next theme. Rather than thinking of the article as a review, think of it as an analysis where the writer has an active role and responsibility in data interpretation and presentation. Some ways to add value to existing data include:

- Highlight things that worked, things that didn't and why that was the case.

- Contrast the best results with the worst and comment on the differences.
- Discuss the strengths and weaknesses of various methods.
- Identify trends and optimums.
- Provide perspective and hypothesis supported by scientific theory.
- Invent novel ways to visualise existing data.
- Generate new figures and tables cumulating data from multiple existing studies.
- Complete statistical analysis or modelling on existing data.

The number of literature references to include in a review article, as with thesis reviews, depends on the research area and objective of the review. If the focus of the review is recent developments in an area, for example, then the number of references will be different from a review encompassing a broader time spectrum. Of course, it can also be that a review focussing on only very recent literature in a popular research area contains more references than a review encompassing the entire history of another more niche research area. Nonetheless, the reference count should be sufficiently high to justify a review's publication and the attention of its readership. Prestigious journals sometimes cap references to 100, excluding in-text weblinks. This might be a good number to aim for although reference counts of 150-300 in a well written article can be more worthy of the reader's time.

WRITING A LITERATURE REVIEW

How best to go about starting a literature review depends on the level of experience the author has with the topic but extensive reading on the latest developments in the research area is always beneficial. Browse the literature using an indexing service such as Google Scholar (www.scholar.google.com) and try to determine the focus and scope of the review before beginning to write. Reading existing reviews or text-books on similar topics or broad summaries such as those found on Wikipedia pages or non-academic websites through a normal Google search can help to become familiar with the basics of a research area and generate ideas on how to group and structure content, although this material will not be cited in the review. Attempt to define the themes and headings that will make up the review at this point in addition to the most appropriate keywords that can be used to locate suitable documents online.

Once a thorough understanding of the current state of the literature has been achieved, download all documents that will be included in the review based on the identified keywords. Store these documents in the main directory of a folder and generate subdirectory folders, each labelled with a single theme or heading that will be included in the review. After this has been completed, open each document in turn and read the abstract, sorting the documents into folders based on theme or heading as appropriate.

It is important to note that the documents in the main directory can most likely be arranged into folders, and hence themes or headings, in any number of different ways i.e., every document may potentially match with any other document based on an instrument used during analysis, a sample type, a variable, etc. It is for this reason that the focus of the review and appropriate themes should be determined before attempting to group documents. Have a rough idea of what each section of the review will be about before starting so that this step does not get confusing. Some documents may relate to more than one theme and can be copied into multiple folders, as necessary.

Once the documents are sorted, generate a new Microsoft Word document, and create titled sections named after the folders or review themes. Take this opportunity to order the themes with emphasis on logical flow of ideas. This structure may change over the course of the writing process as the content that will be presented in each section becomes clearer. Nonetheless, starting with at least some kind of logical progression in mind is critical to ensure that the final product resembles the planned manuscript and subsequently achieves the target value that was identified to justify its publication or thesis impact.

Moving through all literature documents in each subdirectory folder, paste relevant excerpts from each abstract under the appropriate headings in the Microsoft Word document. Excerpts from an abstract may be relevant to multiple sections of the Word document. Ensure that document references are always noted after each excerpt so that the origin of all content can be easily traced. This can be completed by immediately inserting a full reference using citation management software (see WRITING AN ORIGINAL RESEARCH MANUSCRIPT — REFERENCES — REFERENCE MANAGEMENT SOFTWARE — USING EXPORTABLE CITATIONS) or simply as a short note of the document number, author and year or any other form of referencing that makes sense to the author. These notes can be replaced with full references later.

Moving from section to section within the Word document, related excerpts can then be grouped to form the basis for sentences followed by sentence construction. Although the concept of internal and external comparison and discussion described in WRITING AN ORIGINAL RESEARCH MANUSCRIPT — RESULTS AND DISCUSSION and SCIENTIFIC WRITING TIPS lends itself very naturally to the original research manuscript style, it can also be an efficient and practical method to help build these sentences. Look at the grouped abstract-based notes and formulate new sentences that compare key points relevant to each review section e.g., methods used, results obtained, problems, outcomes, applications, across the different studies (internal comparison). Although the compiled abstracts should serve as a solid basis for writing the literature review, it will of course be necessary to return to the results and discussion sections of the original literature documents to obtain additional information, as appropriate. If aspects of these comparison sentences are unclear or could be improved by the provision of background information, additional depth or explanation then search for other relevant literature and write sentences to supplement the internal comparison with appropriate discussion (external comparison and discussion). Be sure to note the tips provided in IMPROVING REVIEW ARTICLE IMPACT IN JOURNALS during this process.

Although the creation of figures and tables is always completed very early on in the original research manuscript writing process the timing of this step when writing a review is more flexible. This is because compiling, comparing, interpreting and summarising a large number of other studies is often an iterative rather than rigidly planned process. There is really no easy and efficient way to read hundreds of documents and create a fool proof writing plan for a review and it would be a waste of time to attempt it. If you are reading the documents anyway you might as well be trying to figure out their place in the puzzle and moving any relevant content across to your review draft at the same time. The result will be a working draft which is slowly refined to perfection. While looking through literature keep your eyes open for particularly attractive or important figures in open access articles that you might want to reproduce, adapt or redraw in your review (see REUSING REPUBLISHED CONTENT IN A REVIEW). However, it is also important to generate and include your own figures and tables. This can be accomplished using the suggestions provided in FIGURE AND TABLE DESIGN. Creation and inclusion of figures and tables is often the final step in the review writing process after all content has been added and finalised. Read through your review at this point and generate attractive figures and tables that illustrate your key points and with emphasis on the suggestions made in IMPROVING REVIEW ARTICLE IMPACT IN JOURNALS.

Once the sections forming the main body of the literature review are completed, use the tips described in WRITING AN ORIGINAL RESEARCH MANUSCRIPT — CONCLUSION and ABSTRACT to assist in the writing of the journal review article's, thesis literature review's or book chapter's conclusion and abstract. The tips provided in these sections rely on the 'Outline of Results and Discussion' created for an original research manuscript but are easily adapted to a review: instead of relying on the writing plan just consider the advice and examples conceptually i.e., a central message and key points on the state of the literature potentially including strengths, weaknesses, research gaps or opportunities and their impact on science and society for the conclusion and the same but with appropriate background information included for the abstract. The instructions in the WRITING AN ORIGINAL RESEARCH MANUSCRIPT — INTRODUCTION section of this guide can also be adapted to a review introduction: generally speaking, just use the advice relating to 'topic and relevance' unless writing a more structured literature review where methods are relevant e.g., patent analysis. Finally, select a title and keywords for the review using the advice provided in WRITING AN ORIGINAL RESEARCH MANUSCRIPT — SELECTING A TITLE and KEYWORDS.

EXAMPLE - LITERATURE REVIEW PARAGRAPH 'BIODEGRADABILITY OF BALLOONS'

The biggest issue limiting the use of *material* (e.g., foil-covered mylar) in *product* (e.g., balloons) is their *property* (e.g., lack of biodegradability) [*reference*]. *Product* (e.g., mylar balloons) typically comprise *material* (e.g., nylon) coated in a *thickness* layer of *material* (e.g., aluminium) [*reference*]. Neither *material* (e.g., mylar nylon), originally produced for use in the US space program, nor *material* (e.g., aluminium) are *property* (e.g., biodegradable), and can take *period* (e.g., hundreds of years) to decompose in the natural environment [*reference*]. Released *product* (e.g., foil-covered mylar balloons) pose a risk to animals, terrestrial and marine ecosystems due to their *property* (e.g., lack of biodegradability) in addition to high voltage powerlines due to their *property* (e.g., electrical conductivity) [*reference*]. Conversely, *product* (e.g., latex balloons) are derived from *natural sources* (e.g., rubber trees) and are inherently biodegradable over *period* (e.g., 6 months to 4 years), which represents just *fraction* of the period associated with *product* (e.g., foil-coated mylar balloons) [*reference*]. Nonetheless, even *product* (e.g., latex balloons) should be disposed of responsibly as their attractive colour, ribbons and tassels can be mistaken for food and ingested by animals if they enter the natural environment [*reference*]. Recent studies suggest that *product* (e.g., latex balloons) degrade much more slowly when submerged in seawater, retaining their elasticity after as long as *period* (e.g., 12 months) [*reference*]. *Product* (e.g., latex balloons) are also increasingly treated with *compounds* (e.g., ammonia, tetramethyl thiuram disulfide and zinc oxide) to prevent bacterial decomposition in addition to *compounds* (e.g., plasticisers) [*reference*]. These treatments greatly affect their biodegradability and make even *product* (e.g., latex balloons) a considerable threat to the environment [*reference*].

REUSING PUBLISHED CONTENT IN A REVIEW

Whether intended for a thesis, journal or book, a common aspect of reviews is the reproduction of figures used in other literature. Many students source graphs or schematics for their review from previously published books, journal articles or conference papers rather than making their own. There is absolutely nothing wrong with doing this, but it is important to understand the rules and regulations behind this practice.

The most important distinction when it comes to reusing content from literature is whether a figure has been reproduced, adapted or redrawn. A reproduced figure appears in exactly the same form in which it was originally published. An adapted figure appears in a modified form with the addition or subtraction of some elements. Both of these reuse types require permission from the copyright holder, which is typically the publisher of the document and not the authors.

Most large publishers have platforms dedicated to handling requests for reuse of copyrighted content, such as the RightsLink® Copyright Clearance Center (www.copyright.com). Those that do not can usually be approached directly through weblinks found on the webpage containing the copyrighted book or journal article. Reuse rights are usually free for those writing a thesis. However, they can be expensive for those wanting to reuse content in a manuscript that will be published in a journal or book.

Sourcing content from open access articles and redrawing figures can be a good alternative for students wanting to reuse figures from literature in a review that they intend to publish in a journal or book. Open access articles are typically licensed under a Creative Commons license and their content can subsequently be reproduced free of charge, although the source must be attributed. If an article can be accessed without a subscription, then this is a strong indicator that it may be open access. However, open access status should still be checked as some journals periodically offer free access to subscription content. Open access articles are typically designated as such on their first page.

Figures that are redrawn do not require copyright permission. Redrawing a figure means creating a new figure from scratch where the existing figure serves only as inspiration for the form of the new figure. If there is any possibility that a figure has been adapted rather than redrawn, then copyright permission should be sought.

REVIEW AND EDITING

When a manuscript draft is completed, it is typically reviewed and edited by all authors. Since a thesis is a requirement for the completion of a degree, there is only a single author: the student who writes it to qualify for that degree. The thesis should, however, be checked by the student's supervisor to ensure that the content is factually accurate, meets the required standards, has no omissions, and is formatted correctly prior to submission. Manuscripts that will be published in a book, conference documents or journal often have multiple authors who must provide their approval before it can be submitted to the publisher.

REVIEWING THESES PRIOR TO SUBMISSION

Students should check their thesis as carefully as possible before giving it to their supervisor. Thesis corrections are rarely a top priority for busy academics. They often take some time and will only be prolonged by students who do not scrupulously review and edit their thesis before submission. Do everything that can possibly be done alone first.

Before submission ensure that the thesis:

- Uses the correct and latest university template if applicable.
- Is complete and contains all thesis content relevant to the degree type.
- Is formatted to the expected university standard and requirements.

Be sure to review the Using this Guide — Monograph Style Bachelor's, Master's or Doctoral Thesis section of this guide for further ideas on these points.

- Is written as concisely and precisely as possible.

- Is supported by attractive yet simple and intuitive figures and tables.
- Has a clear flow of ideas and obvious message and story.
- Emphasises novel results and their impact.
- Reflects the current state of the literature.

Be sure to review the WRITING AN ORIGINAL RESEARCH MANUSCRIPT and LITERATURE REVIEW MANUSCRIPTS sections of this guide for support on these points.

- Has been proofread and any mistakes in the thesis content or expression addressed.
- Contains a reference list that has been checked for errors and converted to plain text with an editable copy retained.

If the supervisor has to comment on or correct grammatical, formatting, and other minor mistakes in addition to identifying inevitable oversights in the thesis content and suggesting opportunities for improvement, then it is a waste of time for both parties. Students are also often under time pressure to finish their thesis, be it depleting finances, an attractive job offer, opportunity to continue their education or university completion deadlines. Providing a well vetted thesis initially and exhaustively acting on supervisor feedback each time it is returned will reduce the time required for each revision and the number of resubmissions required to obtain the final version.

The dead time between avoidable resubmissions is often an even more important factor than the time required by the supervisor to review the thesis. Most supervisors have many students. Each time a thesis is resubmitted it goes back to the bottom of the pile, and fairly so. If working with a busy supervisor with a long waiting list of students, a lack of comprehensive self-review and editing requiring additional resubmissions could result in weeks, or even months, of dead time as the thesis sits idly on the supervisor's desk or in their email inbox.

REVIEWING PUBLICATIONS WITH MULTIPLE AUTHORS PRIOR TO SUBMISSION

When a journal article, conference paper, book chapter or other publication with multiple authors is submitted to the publisher, it is assumed and sometimes must even be explicitly confirmed that the manuscript has been reviewed and its submission approved by all authors. This review, revision, and approval process among the authors prior to submission can be protracted.

Each contributor has their own unique perspective on the topic, preferences on how the data should be interpreted and presented and suggestions for improvements. It is this rich mix of ideas from people with a range of different skills, backgrounds and degrees of experience that makes collaboration so valuable in generating high quality scientific work. But trying to organise a number of busy academics to review a manuscript draft and implement their revisions can be a logistical challenge.

One author should manage this process. This is typically either the first author who wrote the manuscript or the corresponding author (see PUBLISHING JOURNAL ARTICLES — AUTHORSHIP). Regardless of how this process is completed, setting, and enforcing, deadlines for author revisions is often necessary to keep things moving.

Author Revisions in Series

The easiest way to complete the author review process is to send the manuscript to each author in turn and encourage them to make any desired changes with the 'Track Changes' option from the 'Review' tab of Microsoft Word toggled on. This reduces the work required to implement modifications by the author managing the revision process as each change can be reviewed and implemented by simply right clicking it and accepting the change.

Some supervisors avoid this option in student generated manuscripts out of fear that students will simply use the 'Accept All Changes' option of the 'Accept' menu in the 'Review' tab, hence compromising any educational value that the suggestions might have had. In this case supervisors often use comments instead by selecting the relevant text and choosing 'New Comment' from the 'Review' tab.

Once an author has checked the manuscript, their revisions can be reviewed and implemented by the managing author and sent to the next author. This is repeated until all authors have reviewed the manuscript.

While this process makes a lot of sense and is easy to manage, it can be very slow. Delays in a single author's revisions will disrupt and extend the entire review process. Since the manuscript changes as it moves from author to author, it is not possible for other authors to work ahead of the author who holds the manuscript at any given time. Like a game of Chinese whispers, it is also possible that the changes made by the authors that reviewed the draft early in the process may be considerably altered or skewed over the course of the full review. It is subsequently important for all authors to check the final manuscript again at the end of the review process to ensure that they are happy with the final product.

Author Revisions in Parallel

A faster but potentially more complicated author review process is to send the manuscript to all authors for review at once. Doing so will cut the time required for the review to that of the slowest author, a massive improvement on the cumulative time required for every author to review the manuscript in turn. The degree of difficulty involved in this process will depend on the number of authors and the extent of the revisions that are necessary. It is considerably more work for the author managing the review process. Use of the 'Track Changes' feature by the authors to quickly and directly modify text in Microsoft Word becomes a nightmare when changes from many authors must be consolidated into a single document. This method is subsequently better suited to polished manuscripts with as few authors as possible.

Request that authors make their suggestions as comments where possible and limit in-text changes to critical errors. Once all authors have returned their version of the manuscript draft use either the most extensively modified version or the original document that was sent to the authors as the new revised manuscript. Open each individual author's version of the document and the new revised manuscript, arranging them next to each other in windows of the same size and scale.

If multiple computer monitors are available, then ideally use several monitors to view the new revised manuscript and all author versions at once. If not then use a single split monitor displaying the new revised manuscript on one side of the screen and an author's revisions on the other, toggling between the different author versions as appropriate. Scroll all manuscripts together, moving each of them in turn to ensure that they are all kept in more or less the same position within the document. Compare the comments from the various authors, considering the suggestions, supporting, and conflicting points before making the appropriate changes to the new revised manuscript and then scrolling further across all documents.

There can be some degree of work duplication using this method as multiple authors may make similar comments, an occurrence that would have been avoided if the authors to first review the manuscript had made these changes before it was sent to the next author in the queue. However, the parallel review process does promote originality, with each author potentially offering slightly different perspective on the same suggestion. It also helps distinguish more important, or at least more obvious, revisions from the others depending on the frequency with which a change is suggested.

Regardless of review methodology, or how comprehensively co-author feedback is implemented, the final revised manuscript must always be checked once more by all authors to ensure that they are satisfied with the version that will be submitted to the publisher. A hybrid approach to the review process is also possible, using both series and parallel review methods as appropriate at various stages.

SUBMITTING A THESIS

Although often the most emphasised component of a degree, the submission of a thesis isn't usually the only requirement for completion of a university programme. Every university has its own thesis submission protocols and procedures but generally speaking thesis submission is the final step and cannot be completed until all university defined programme requirements have been met. These may include:

- Compulsory programme specific coursework or exams.
- Generic coursework or exams that meet an expected credit aggregate requirement.
- Annual progress reports, presentations, or assessments.
- A certain number and type of publications for doctorates, potentially as first author.
- Settlement of all outstanding financial obligations to the university.

Students should familiarise themselves with the thesis submission process at their respective universities well before they start preparing their thesis. There is sometimes considerable paperwork involved in registering a thesis submission, selecting examiners and if applicable setting a date for a public thesis defence, all of which must be approved by supervisors, administrative staff and programme or departmental managers. Be sure to understand these processes and the time required to complete them, which can potentially be months, and to factor it into the thesis completion timeline.

BACHELOR'S PROGRAMMES

A bachelor's thesis is usually examined by the student's supervisor. Following submission, typically electronically via email, the supervisor will review the thesis and either immediately assign it a grade or provide feedback and revisions that the student should implement before the examination can be finalised. Several iterations of feedback and revisions may be necessary prior to final acceptance by the supervisor.

The thesis may sometimes be cross examined by other academics within the same university, but this is usually the extent of the examination process. Once the thesis has been passed the degree may be awarded provided all other programme requirements have been met. Bachelor's theses may be archived by the university library, but this is not always the case. They may instead be archived by the supervisor.

MASTER'S AND DOCTORAL PROGRAMMES

Once a master's or doctoral thesis is finished and has been reviewed, edited, and approved for submission by the supervisor, it is typically submitted to the university administration. Both a digital copy and several hard copies are usually required for examiner use and archiving purposes. The thesis is often checked by the university administration to ensure that it is formatted to university standards and does not contain plagiarism using commercial software before being forwarded to the approved examiners.

Thesis Examination

Thesis examiners are usually recommended by the student or their supervisor but must also be approved by the university administration. Various institutions have differing requirements for master's and doctoral thesis examiners, but they are usually internationally recognised senior academics such as professors. They may reside and work in a country other than the country in which the thesis was completed. The main requirements of an examiner are a scientific reputation in line with university expectations, considerable expertise in the thesis topic and a lack of any conflicts of interest with the student and supervisor.

Thesis examinations form a very small part of an academic's job and are often not a top priority. Academics are under no obligation to accept a thesis examination request and there is little financial incentive to do so. A token payment is generally provided per thesis examination, but such a sum would very rarely justify the time required to complete such an assessment. Thesis examination is rather a service to the scientific community that most academics take seriously but that can take quite an extended period (sometimes weeks but often months) to complete.

Once the examiners have read the thesis, they will usually write a report outlining their assessment of its content. Formatting generally doesn't receive much emphasis in this report, unless major errors exist, since a fair degree of flexibility is provided to students in thesis formatting and any specific requirements are usually enforced by the university administration and supervisor before the examiners receive it. Instead, examiners are generally interested in the thesis content and if it meets the expectations and requirements for award of the degree. This may include ensuring that:

- The thesis is complete, i.e., contains all necessary elements.
- The thesis content is factually accurate.
- The thesis content is comprehensive and without omissions.
- Data is presented clearly and in an appropriate format.
- The adopted experimental and analytical methodologies are sound.
- There is a clear and obvious novel contribution to science.
- Participation in scientific dialogue is demonstrated e.g., publishing, conferences.

Thesis examiners are usually required to assign a numerical and word-based grade to the thesis and provide comments. They may also suggest revisions that should be implemented by the student before the thesis can be passed. The type and extent of feedback provided by an examiner will vary from person to person. Some examiners may home in on specific aspects of the thesis with which they are especially experienced, such as statistics or particular experimental methodologies, while others may provide more general feedback. It is these differences in individual interpretations, skills, and degrees of experience that make the allocation of multiple (usually two) examiners necessary. A pass without revisions is also possible but is less common.

Student Revisions

In some countries, no revisions are allowed to a thesis after it has been submitted. It is simply passed or failed by the examiners based on the content that it contains when they receive it. However, in most cases examiners can request minor or major revisions to the thesis. Students must then carefully consider the examiner's feedback and make appropriate changes before resubmitting the thesis. Minor and major are somewhat subjective terms in this context and different examiners have different ideas of what these descriptions mean. Minor revisions could be errors in the thesis text, figures or tables that need to be corrected while major revisions may be text-based but could also indicate fundamental errors, oversights or omissions in conceptual understanding, experimental methodology, interpretation of results or similar.

Response to Examiners' Comments

Students must usually provide a response to every comment made by the examiners upon resubmission. As time consuming as addressing examiner comments can be, insight and perspective from other experts in the field can radically improve a thesis. Students should remember that examiners' comments are aimed at helping them and safeguarding the quality and integrity of the research area. Try to be as open as possible to criticism. Consider every comment carefully and, if possible, implement it. Examiners often have valuable suggestions, feedback, and perspective.

A response to the examiners' comments detailing changes made to the revised thesis should be compiled in a separate document. This document has no special formatting requirements and for many is neatly arranged but fairly plain. To prepare this response paste all remarks, queries, and suggested changes into a Microsoft Word document. Break every individual entry from each examiner into a separate paragraph so that they can be addressed point-by-point. Every thematically different element of each remark, query or suggestion should be separated to ensure a clear and focussed response can be made to every aspect of every entry.

It is not compulsory to make every change that an examiner has requested. If there is a good reason to justify not making a requested change e.g., the examiner has misunderstood an element of the thesis or has touched on a controversial topic, it is acceptable to rebut the comment. However, if a requested revision cannot or should not be made, then at least attempt to clarify the potential for misunderstandings in the thesis text and reinforce the original statement with additional context, relevant experimental work, or references. It should be apparent that the comment has been considered at length and every attempt has been made to address it in the most appropriate way. In general aim to implement all examiner feedback unless there is a good reason not to and, if so, be sure to state and discuss this reason in detail in the response to examiners' comments document.

Structure the response document using appropriate labels such as:

- 'Examiner comments' arranged by examiner number and broken into individually labelled 'examiner queries' which are then addressed with an 'author's response'.
- A detailed 'author's response' to each examiner query that comprehensively describes the changes made in the revised thesis based on the query or provides additional information or justification as to why no change has been made.
- 'References' to support discussion or rebuttal of examiner queries. If an examiner requested change has not been implemented, a justification backed by references is advisable. If additional references were explicitly requested by the examiners for any reason, they should also be included.

Revised Thesis

Revisions should be made in a revised thesis document. Create a new copy of the original thesis and retitle it to reflect its revised nature. Work through the comments one at a time, making the appropriate changes in the new document and then providing a detailed description of these changes in the response to examiners' comments document. It is important to note that universities do not usually accept the use of track changes or other Microsoft Word review tools for revisions. Implement revisions in the manuscript body as plain text instead.

Once all changes have been made, resubmit the revised thesis and response to examiners' comments document to the university administration. The revisions will be reviewed and if satisfactorily completed the thesis will be passed. At some universities this signifies the end of the assessment process and, provided all other programme requirements have been met, the degree can be awarded.

Public Thesis Defence

Other universities (especially European ones) require a public defence (viva) of the thesis. This normally comprises a presentation of the thesis content by the student followed by a question session during which the student must justify their methodologies and design of experiment and defend their results, discussion, and conclusions. This presentation is attended by the thesis supervisor, examiners, and an independent university representative (chair) and is open to the public.

Once the thesis has been examined and passed, a date can be scheduled for the public defence. This is generally arranged by the student or thesis supervisor who negotiates with the thesis examiners and university to establish a suitable date. Public defences generally take place physically on university grounds but in some cases can also be performed online using video conferencing. Once a date has been selected, it is usually advertised to the public through a university bulletin or similar.

Public defence protocols differ by university but usually there is a fairly short timeslot allocated to the student's presentation (20-30 mins). The presentation subsequently must be very concise and focus only on the most important aspects of the thesis. A more extended dialogue follows with the supervisor and thesis examiners permitted to engage with the student one at a time with questions on the thesis content, which the student must answer. These may relate to:

- Experimental methodology selection and appropriateness.
- Scientific theory behind the function of analytical instruments or procedures.
- Theory and phenomena behind any results or discussion or more broadly any aspect of the thesis topic.
- Rationale for decision making in hypothetical scenarios.
- Substantiation of results, discussion, or claims.
- Perspective on the future of the research topic.

Although rough time restrictions are allocated to examiner questioning, they are generally loosely enforced, and this part of the defence can take quite some time. The dialogue between a student and the examiners is subsequently often the longest part of the defence. Once the examiners have finished their questioning, the floor is opened to the audience for questions from the general public. This entire process is managed by the defence chair, an academic who generally has no prior association to the student or thesis and is assigned to ensure that the defence conforms to university protocols. Public defences generally last between 1.5-2 h, although considerably longer defences are also possible.

Thesis Archiving

Once a thesis has been passed and a public defence completed if applicable an electronic and hard copy of the thesis is usually archived by the university and sometimes state or national libraries. Most master's and doctoral theses can be accessed online for free by the general public through a university library repository. Indexing is performed based on a student-supplied title, abstract and keywords.

PUBLISHING JOURNAL ARTICLES

SELECTING A JOURNAL

Selecting which journal to submit a manuscript to can be a difficult decision. After all, the time required to format a manuscript to journal requirements and produce submission supplements, such as a cover letter, graphical abstract and highlights, cannot be discounted and may be very quickly rendered redundant by a rapid desk rejection from a journal editor. Nonetheless, the effort required at the submission point of the publishing process only represents a tiny fraction of the overall time invested in producing the article and it is worth aiming as high as realistically possible when selecting a journal. There is nothing to lose by trying to publish a manuscript in a higher quality journal, with rejection often very prompt and the opportunity to try another journal later always present. An exception to this concept exists in highly competitive time-sensitive research areas where multiple research groups are potentially working on the same topic and being the first to publish is important.

Regardless of which journal will be targeted, spend a little time researching its reputation, subject area, and readership prior to submission. The Scimago Journal and Country Rank website (www.scimagojr.com) allows authors to search for and view key metrics on most journals and can be a good resource for establishing their reputation. It also provides journal rankings by subject area, category, or region (www.scimagojr.com/journalrank.php) and can help to identify suitable journals for specific research areas. Once a journal has been selected, check its scope to ensure that it matches the manuscript theme. That said, journal scopes can be quite broad, and the suitability of manuscripts is ultimately based on the opinion of the editor who checks them. If the scope and all other submission requirements are met, then proceed to format the manuscript to the journal requirements.

AUTHORSHIP

Publication output is very important in the academic space as it is often linked with promotions, academic and university reputations, and funding both within the university and national and international funding bodies. Authorship is subsequently a sensitive topic among academics. If someone contributes to a manuscript but rather than being listed as an author is noted in the acknowledgements or left out entirely, then they will receive no recognition for their work and no citations if the manuscript is referenced in the future. Academics are subsequently very keen to be listed as a co-author on manuscripts that they contribute to.

University academic and technical staff usually do not receive anything for assisting with a scientific study or manuscript. Of course, they receive a salary for working at the university but the requests that are made of them sometimes fall well outside their job description and the boundaries of traditional compensation. These staff often help out of interest in the topic or purely good will. They are often overworked and also rely on recognition of their efforts for internal appointments and even funding of their departments, which is commonly at least implicitly linked with publication output.

When considering whether to include someone as an author on a manuscript remember the fact that it does not cost anything to add an extra name to the paper and the person whose name is being left off will get nothing for their time and effort if they are not included.

It is also important to consider contributions with their relevant context. Substantial contributions can be trivialised in some people's minds based on a lack of consideration for proportionality. The hour that a professor takes to make revisions on a manuscript may not seem as substantial as some other contributions on face value, but they may have 20 years of experience in the research area. How long would it take someone else to revise the manuscript with the same degree of precision and insight? The hour spent with the technician as they analyse a sample with a million-dollar instrument and explain the results might not seem that substantial but perhaps those results were the most important data in the entire manuscript.

People like to be recognised for their work. If someone who made a meaningful contribution to a manuscript is not included on it as a co-author, they may be less than enthusiastic to get involved again in the future. Whereas academics who see their time and effort appreciated and acknowledged by others are often excited to provide additional help and support and may even initiate new collaborations in return.

Of course, it is also possible to go too far when it comes to including people as co-authors on a manuscript. Technicians for example generally do not qualify for authorship unless they play an active role in the analysis, interpretation, and discussion of results. Credit must be fairly distributed; however, every additional author does inevitably dilute the perceived value of the other contributions. Use the CRediT Contributor Roles Taxonomy as a guide when trying to establish whether or not someone qualifies for authorship on a manuscript (www.elsevier.com/authors/policies-and-guidelines/credit-author-statement).

Many journals require manuscript submissions to be accompanied by a CRediT-based author contribution statement. This is a separate document with each author's contribution to the manuscript summarised using a few words based on the contribution types described in CRediT. If a person has contributed to the manuscript but does not meet the requirements for inclusion as an author as outlined in CRediT, be sure to include a brief description of this person's role in the study in the acknowledgements section at the end of the manuscript instead.

Once the authorship of the manuscript has been determined, write the author list as a row of names beneath the title, including both given and family names for each author. Inclusion of initials for middle or additional names is optional but can help to distinguish authors who have popular names from others with similar names. Commas should be used to separate author names with the word 'and' or an ampersand (&) directly preceding the last author's name. The author list may spill across multiple lines if necessary.

It is important to note that, although any citations that the manuscript receives will be allocated to all authors in the same way, the order of the names in the author list is still important. It is generally recommended that authorship order be discussed and agreed upon prior to beginning experimental work to avoid conflict later. This is not always possible and author roles can change over the course of the study as unexpected challenges and complications arise.

First Author

The first author of a scientific manuscript is the author whose name appears in the first position of the author list. This is arguably the most desirable authorship position as it is associated with the most substantial contribution to the manuscript.

First authors should have personally performed the majority of the work involved in the completion of the scientific study and manuscript. Contribution types for this position can vary but often include active participation in all aspects of the study including its conception, design of experiment, and experimental work. Critically, first authors should play a pivotal role in the interpretation of results and writing of the first manuscript draft.

Holding first authorship positions on publications is especially important when completing a doctorate. This is because doctoral programmes constitute scientific training with competencies that must be established prior to the award of the degree. Doctoral students must prove that they are capable of all fundamental elements of scientific practice to pass. For publications to be credited towards their scientific training, it is subsequently important that the student performed the majority of the study and manuscript preparation themselves.

Although not especially common, it is also worth noting that authors can share the first authorship position (or any other position for that matter, although their less prestigious nature makes this practice less relevant). Of course, one author must still be listed first in the author list; however, if desired an asterisk (*) can be added to the first two names and a footnote made stating that the authors contributed equally.

Co-authors

Anyone who made a meaningful contribution to a scientific study, or the drafting of its manuscript as described in CRediT, should be listed as a co-author. Co-authors occupy the authorship positions following the first author and should be listed based on the extent of their contribution, starting with the most substantial and ending with the least. Judging contribution weight can be challenging across different and not necessarily easily comparable types of contributions. However, most academics if already excluded from a first, last or corresponding authorship position will generally not care much where in the order their name appears.

Last Author

The last authorship position in the list is normally associated with a conception and supervisory role. The person in this position typically came up with the idea for the scientific study, may have attained funding enabling the experimental work to be completed and supervised the process. For doctoral students, it is typically the position that the supervising professor will take.

Corresponding Author

The corresponding author is the author responsible for communication with the journal during the manuscript submission, peer review, and publication process. They must typically prepare, complete, and manage the submission including ensuring the manuscript meets the journal's administrative requirements, writing a cover letter to accompany the submission, communicating with the editor, preparing a rebuttal during peer review and handling all licensing issues upon publication including conflicts of interest paperwork, copyright transfer and open access options. The corresponding author is also responsible for all manuscript-related communication with the scientific community and general public after it has been published.

There are many reasons why readers or the wider scientific community may want to contact the authors of an article after it is published. These include questions regarding the methodology used in the article to allow others to replicate the work, discussion or comment on the work, requests for access to the original experimental data, expressions of interest in collaboration and media enquiries.

Corresponding authors are generally marked with an asterisk in the author list and their contact details, including an email and physical address, supplied below the affiliations or in a footnote. The supervisor or project leader is often the corresponding author, as they tend to have experience in manuscript submission and management and are permanently engaged with both the research area and an institution. The last and corresponding author roles are subsequently often held by the same person. That said, any author in any authorship position can potentially hold this role. Multiple corresponding authors can also be assigned.

Ideally the corresponding author will have fixed and permanent contact details, which are preferably institutional but could also be personal, that will still be in use many years after an article has been published. This is to facilitate lasting access to an author who will represent the article and be contactable long after it has been written, and other authors have potentially changed institutions or career paths.

The corresponding author should be familiar with all aspects of the study and have archived copies of all data sets used in the manuscript. This role, while bearing the weight of responsibility for future representation of the article and correspondence, is often desirable as benefits, such as collaboration opportunities and media coverage, are usually directed to the person occupying this position.

AFFILIATIONS

An affiliation details the formal connection between an author and their research institution. Although universities are common affiliations, it should be noted that independent research institutions and companies are just as suitable. All that is really required for an affiliation is a registered entity and a mailing address.

Use of university affiliations is not limited to university employees. Bachelor's, master's, and doctoral degree students and even alumni are usually able to use an appropriate university affiliation e.g., the department that they work or study in or previously worked or studied in. If in doubt it is always possible to email the university to seek permission to use their affiliation. However, since internal and external funding often depend on publication output that is in turn linked with affiliations, most universities and departments actively encourage use of their affiliation.

Students and researchers often move institutions, a fact which has no effect on previous affiliations. Published literature will reflect an author's affiliation at the time of the study. If they move to a new institution, they will simply adopt the new affiliation for any research published following the shift. If research work spans a duration over which an author worked at more than one institution e.g., over the course of a research visit to another institution or for extended projects completed over several years at several different universities, then multiple affiliations can be listed on a manuscript resulting from this work.

Affiliations appear beneath the author list or in a footnote as a continuous list of all institutions affiliated with the manuscript authors. They start with the department, institute, or school name, followed by the university or institution name and the mailing address of the building occupied by the relevant department. Do not list the generic mailing address of the university.

Each affiliation is labelled with a letter or number in superscript and starts on a new line with the potential to span multiple lines if necessary. These letters or numbers correlate with identical entries appearing next to the relevant author's name in the author list. They move in an ascending fashion with their location in the author list dictating the order of the affiliations list.

EXAMPLE - AUTHOR LIST AND AFFILIATIONS

Adam A. Appleby,[a] Benjamin B. Barrow[b,c] and Charlotte C. Crown[a,d,*]

[a] *School, Department or Institute Name, Faculty Name, University Name, Campus Name, Street or Mailing Address, Postcode, City, State, Country*

[b] *School, Department or Institute Name, Faculty Name, University Name, Campus Name, Street or Mailing Address, Postcode, City, State, Country*

[c] *School, Department or Institute Name, Faculty Name, University Name, Campus Name, Street or Mailing Address, Postcode, City, State, Country*

[d] *School, Department or Institute Name, Faculty Name, University Name, Campus Name, Street or Mailing Address, Postcode, City, State, Country*

* Corresponding Author: *Email address.*

FORMATTING A MANUSCRIPT FOR SUBMISSION

All journals have expectations on how manuscript submissions should be structured, formatted and what sort of content they should contain. These standards are in place to streamline the editorial process and maintain consistency in the presentation of publications across the same journal. The expected format for manuscript submissions to any given journal is normally detailed in a document known as the 'Guide for Authors', which is usually available as a webpage or downloadable PDF on the journal's website.

Using the Guide for Authors

It is critical to thoroughly review the target journal's Guide for Authors to ensure that manuscripts are prepared to their guidelines. Failure to do so may result in immediate rejection during initial screening by the editor. The Guide for Authors document can vary considerably by journal, even for journals from the same publisher. That said, publishers do often have similar manuscript formatting and content requirements across related journals. Ensure that the Guide for Authors being consulted during manuscript formatting is for the correct journal.

Understanding Journal Expectations

In addition to detailed manuscript formatting requirements, the Guide for Authors often also provides general information about the journal that can be advantageous in improving a manuscript's chance of acceptance. It may for example detail the types of articles accepted by the journal (see Using this Guide — Journal Articles — Common Journal Article Types), the requirements for these articles, the journal scope, included and excluded topics and audience. Use this information to understand:

- The journal's expectations for any given article type.
- What they consider suitable and interesting content for their journal.
- What sorts of people are going to be reading their articles.

Ensure that submitted manuscripts conspicuously conform to these expectations to improve the chances of a successful initial screening by the editor. It is worth looking at previously published content on the journal's website for examples of accepted manuscripts that meet these expectations. The (open access) article licencing options and journal impact factor provided in the Guide for Authors may also be of interest when estimating the impact of a future publication. Licensing options in particular should be thoroughly checked during the journal selection process by authors who are not prepared to pay publication fees associated with open access licensing. Some journals have a compulsory open access licensing structure and authors will be required to pay for such licensing upon acceptance of their manuscript before it will be published (see Using this Guide — Journal Articles — Journal Article Access and Licensing).

Journal Abstracting and Indexing

The Guide for Authors will generally provide an overview of the abstracting and indexing services in which the journal is included. The inclusion of journals in these services is important for academics since researcher reputation is often represented using online service-specific profiles e.g., Web of Science, Scopus, Google Scholar, which track publication output and citation. Published content appearing in journals that are not indexed by a particular service will not be included in the author's profile on that service. Since university and funding body metrics are often based on these indexing services, such omissions are highly undesirable. Reputable journals are generally indexed by Web of Science and Scopus. If a journal is not indexed by these services, then approach it with caution. Google Scholar indexes everything including webpages. For more information on indexing and its relevance to authors, see the USING THIS GUIDE — INDEXING AND AUTHOR PROFILES section of this guide.

Getting to Know the Journal Editors

The identity and expertise of the editor assigned to assess and manage a manuscript submission to a journal at least initially depends on who the submission is addressed to. While the convention is to address the submission to the editor-in-chief, authors sometimes have the opportunity to explicitly suggest which editor should manage their manuscript during the submission process.

Editors are typically listed in the Guide for Authors and on the journal's website along with their areas of expertise. It can be advantageous to review the journal editors prior to submission and if possible, select the most appropriate editor for any given manuscript submission to ensure that any subtle or niche value that it may contain does not go unnoticed by a differently experienced editor. Otherwise, address the submission to the editor-in-chief of the journal who will then process it themselves or delegate it to an appropriate editor at their own discretion.

Flexibility in Formatting of Initial Submissions

For some journals, the Guide for Authors is a very comprehensive document with detailed requirements, while for others it can be quite short and vague. Several large publishers, e.g., Elsevier, are growing more tolerant when it comes to initial formatting requirements, postponing compliance with journal format until after the peer review process and allowing authors greater flexibility at the time of the initial submission. This is useful as it makes the time-consuming manuscript formatting process less laborious for initial submissions, which can potentially be rejected on the same day that they are submitted. The formatting time investment instead occurs during the revision stage, at which point the manuscript is more likely to be accepted for publication. Other publishers e.g., Multidisciplinary Digital Publishing Institute (MDPI) provide fairly self-explanatory Microsoft Word templates for initial submissions, which make the manuscript formatting process very intuitive. While traditional manuscripts are very plainly formatted, this template often aesthetically resembles the final product that will be published in the journal.

Traditional Page Formatting Requirements

Traditionally formatted manuscripts are very plain. They typically utilise fonts such as Times New Roman or Arial and wide line spacing, e.g., double spacing and wide e.g., 3 cm margins. Full justification is not typically used for the manuscript text i.e., text does not have a continuous right-hand margin. Use left aligned text selected from the 'Home' tab in Microsoft Word instead.

Line numbers are used on every line of the entire document without exception and the pages are also numbered. Line numbers can be activated in Microsoft Word by selecting 'Line Numbers' and then 'Continuous' from the 'Layout' tab. Page numbering can be activated by selecting 'Page Number' followed by the desired page number location and style from the 'Insert' tab.

All sections, excluding the acknowledgements and references, are numbered with Arabic numbers and their subsections labelled utilising decimal system variants. Footnotes are only used where necessary and are formatted using the default settings in the word processor.

Most journals use a unique referencing system that is described using examples in the Guide for Authors. Use reference management software and a downloaded journal style template to quickly format references to the journal standard (see WRITING AN ORIGINAL RESEARCH MANUSCRIPT — REFERENCES).

Although not usually specified in the Guide for Authors, journals also typically assume that:

- No Microsoft Word review tools, such as track changes or comments have been used in the submitted manuscript (except for publishers that explicitly request revisions be made in this way, such as MDPI).
- All references generated using reference management software or other field codes have been converted to plain text.

Content Requirements

Although journals are increasingly tolerant of variations in page formatting and referencing for initial submissions, the same leniency should not be assumed when it comes to content requirements. These are also outlined in the Guide for Authors.

Required content for a successful submission is often stipulated in addition to the order in which it should appear e.g., title, authors, affiliations, abstract, keywords, main text, acknowledgements, appendix, references, figure captions and then tables.

Page or word limits may also apply and may vary by article type. If a page limit is imposed, then text size and spacing is typically specified and must be adhered to. Page limits usually include figures and tables but not references. A word limit is also often applicable to abstracts, sometimes accompanied by content suggestions.

Keywords are typically restricted to 5-6 in number. Plural forms, abbreviations and terms which cover multiple concepts are usually restricted. Some journals provide keyword pools from which suitable keywords must be selected while others allow authors to choose keywords more freely.

Figure and table limits are sometimes expressed as a combined total or can comprise specific restrictions on one, the other, or both. Some journals request that figures and tables are distributed throughout the manuscript in their appropriate positions, while others stipulate that they should be grouped at the end of the manuscript or submitted as separate files. This makes little difference to anyone other than the reviewers as figure and table layout in the final journal article is determined by professional editorial staff during the production process.

Tables and equations should always be presented as editable text (with equations in line with the normal text where possible) and never as images. Use of vertical lines and shading in tables is also ill advised.

Captions should generally comprise brief titles and explain all symbols and abbreviations used in the figure or table. Some journals request that captions are grouped as a list at the end of the document while others prefer that they accompany the figures and tables throughout the manuscript.

Most journals recommend the use of SI units (International System of Units).

Some journals have a limit on the number of references that may be included in a manuscript, especially for review articles. Weblinks inserted in the main text body are typically excluded from the reference list and hence not included in the count.

Electronic Artwork Requirements

Most journals impose requirements on electronic artwork e.g., figures submitted with the manuscript to ensure that the files that they receive are suitable for publication purposes. These requirements typically relate to:

- The font used in electronic artwork. Arial, Helvetica, Times New Roman, Times, Symbol or Courier are normally recommended.
- The resolution, maximum size, and preferred formats for electronic artwork. This varies by file type and should be checked in the Guide for Authors. Preferred formats include TIFF, JPEG, EPS, and PDF. Files optimised for screen use, such as GIF, BMP, PICT, and WPG files are not typically recommended.

Journals increasingly aim to promote more interactive and impactful data presentation methods in their publications. Subsequently, most welcome submission of electronic annexes, such as short videos, computer-enhanced images, audio clips and large databases alongside manuscripts. Inclusion of such elements is, however, not compulsory.

Data Availability Requirements

The availability of the experimental and analytical data on which a manuscript is based is a key aspect of transparency and integrity in science. Aside from the obvious added credibility that a manuscript supported by readily available datasets enjoys, making data available to other scientists can help them to better understand manuscript content, reproduce the experimental work and build upon it more rapidly in their own experiments.

Manuscripts focus on interpretation and discussion of results that fit a given scope and story. Only a percentage of the data collected during the study is likely to be relevant enough to the manuscript key points to be included in the figures, tables and discussion. The remaining data, however, while perhaps not applicable enough to be covered in any degree of detail in the manuscript, can potentially be exceptionally valuable to researchers working within a different scope. They may use the excess data as a reference or point of comparison for their own studies or may even be able to derive new meaning from it for a different application or theme.

The best place to store data in an open format that enables easy access and use by other researchers is a suitable repository. Many repositories are available and are generally discipline specific although generalist repositories for scientific data also exist. In general, it is advantageous to use a discipline specific repository since they are more experienced in handling data relevant to a particular field and may subsequently have provisions in place to catalogue and manage certain data types more efficiently. The target audience of the manuscript is also often more familiar and comfortable with repositories specifically designed and managed for their research area. Uploading data to a repository is normally a simple and free process completed through the repository website.

Manuscripts can sometimes be linked with uploaded data during submission using the database identifier code. Otherwise, details on the location and availability of research data required to reproduce the work undertaken in a study can be included in a data availability statement in the manuscript. Database identifiers, weblinks or instructions on attaining relevant data from the manuscript authors can be included in this statement.

Data must at the very least be available from the authors upon request. Researchers must always be able to substantiate their work, even well after publication. However, there are also plenty of reasons why scientists may be reluctant to make their data widely available to the general public including legal, ethical, technical concerns or also for example because the data forms part of an ongoing study. If any of these are applicable, then this should be noted in the data availability statement.

Some publishers, such as Elsevier, also provide options for publishing data- or methods-only articles e.g., Data in Brief and MethodsX. Such articles are reviewed, formatted, and indexed. However, their open access format often constitutes an unjustifiable financial burden for authors. With free scientific repositories available for data storage, paying for such services is often unnecessary.

Less Traditional Content Requirements

While many journals have similar content requirements and structures, others explore interesting variations and extensions of existing methods of scientific communication. These could include examples such as:

- Breakout style text boxes adjacent to or within the manuscript text, where explanations of basic concepts or theories, detailed mechanisms or discussion of case studies can be provided, bringing otherwise lost accessibility and context to the article without impacting the word count or main flow of ideas.
- A distinct outstanding questions box to conspicuously highlight important questions for future research. These may be expressed using bullet points and provide input and guidance on new directions for the field rather than blending this content into the main text body.
- Relocation of the 'Materials and Methods' section to the end of the manuscript or even to the supplementary material to deliberately emphasise the other more important and engaging aspects of the manuscript.
- Designation of a key figure or key table as opposed to a graphical abstract. This sits within the manuscript body and provides a visual summary of the most important aspects of the article in a clear, didactic, and impactful way.

Certain journals may also have requirements on the type of content appearing in traditional manuscript sections. These may include examples such as:

- Limits on reference scope to focus only on very recent developments (2-4 years).

- An emphasis on critical analysis of other scientific studies, strong perspective on future developments, challenges, experiments, and methods (especially for reviews).
- Hypotheses, author opinion and comment.
- Overly simplified manuscript text and captions that appeal to a broad audience.
- Compilation and presentation of large quantities of tabulated data.

Be sure to read the Guide for Authors properly and understand any special requirements that a journal may have prior to manuscript submission.

Common Submission Supplements

Supplementary documents are often required when submitting a journal article. Common submission supplements are discussed in the sections below.

Graphical Abstract

A graphical abstract is an image that is explicitly designed to convey the essence or core aspects of a manuscript in a similar fashion to a traditional text-based abstract (FIGURE 19). These images are increasingly becoming a mandatory supplement for manuscript submissions in many journals. Aspect ratio, dimensions and file format requirements vary by journal and are usually detailed in the Guide for Authors. Microsoft Word or PowerPoint are often useful tools for designing graphical abstracts for authors without considerable graphic design expertise.

The page or slide size should be first set to the journal graphical abstract dimensioning requirements. In Microsoft Word this is achieved by selecting 'Size' from the 'Layout' tab then 'More Paper Sizes' followed by entry of the custom dimensions. Select 'Margins' followed by 'Custom Margins' from the 'Layout' tab and set the paper margins to zero before doing so to avoid error messages. In Microsoft PowerPoint select 'Slide Size' from the 'Design' tab then 'Custom Slide Size' followed by the desired dimensions. Once this has been achieved use the tips in the WRITING AN ORIGINAL RESEARCH MANUSCRIPT — FIGURE AND TABLE DESIGN

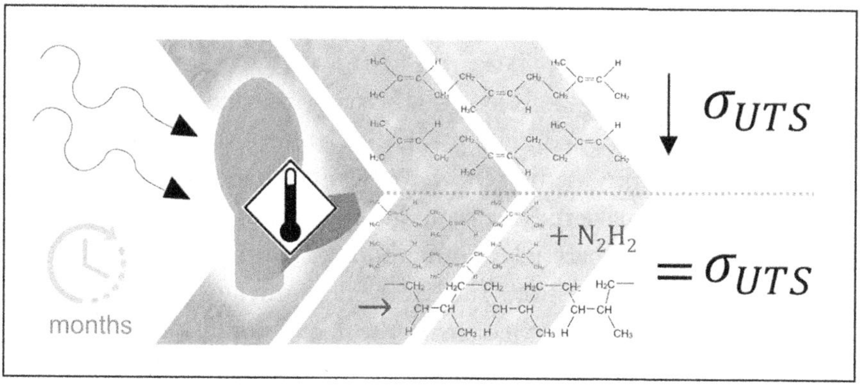

FIGURE 19. *A sample graphical abstract for a study investigating the effect of thermal aging on the tensile strength (σ_{UTS}) of untreated and diimide (N_2H_2) treated (hydrogenated) natural rubber balloons. Moving from left to right, the first third introduces the concept of the study, the second third the most important variable and the final third the results. Use of space and simplicity is maximised. Only the most critical elements of the message and theme are represented. Use of text and representations of data is minimised. Colours appearing in this figure may not be interpretable on some devices and in some print formats.*

— SCHEMATICS section of this guide to produce an attractive graphical abstract.

There is a lot of freedom of choice when it comes to designing graphical abstracts, but the following points are worth considering:

- Maximise the use of the space available. Do not leave large empty spaces at the image edges, around or between the various visual elements.
- Keep the graphical abstract as simple as possible. Minimise the number of visual elements and maximise their size.
- Ensure that the visual elements and the overall impression of the graphical abstract match the most important content, 'central message', and theme of the manuscript as closely as possible.
- Minimise the use of text. If text must be used, then ensure it is as large as possible.
- Avoid including graphs. They are usually visually unappealing, hard to read and their relevance to the manuscript is not usually clear at a glance. Use illustrations or symbols to convey

the intended meaning instead e.g., an arrow pointing up accompanied by a symbol associated with a parameter.

When conceptualising a graphical abstract, it can help to split it into thirds based on the key content of a traditional text-based abstract. Moving from left to right, the first third can be used to introduce the context of the research or make the first point in a logical progression. The second third can detail the materials and methods or continue with the second point in a logical progression. The last third can then be used to illustrate the key results of the research or make the final point in a logical progression.

Highlights

A series of short bullet points that convey the key results of the manuscript and outline any novel methodologies used in the work is a mandatory supplement for original research manuscript submissions in many journals. These so-called highlights assist search engines in delivering relevant content to the target audience and provide readers with additional succinct insight into an article.

Typically, 3-5 in number, these bullet points must be presented in a separate Microsoft Word document and usually cannot exceed 85 characters each in length (including spaces). Use of relevant topic-specific keywords should be promoted in the text but jargon, abbreviations and acronyms should be avoided. Try to maintain a logical flow from point to point, providing the key results of the manuscript without attempting to cover all manuscript content.

EXAMPLE - HIGHLIGHTS

- The aging behaviour of natural rubber balloons is highly temperature dependent.
- Balloons aged at ambient conditions do not exhibit cracking and reduced strength.

- Thermal aging rapidly embrittles untreated balloons rendering them unfit for purpose.
- Diimide-based rubber hydrogenation mitigates temperature-related property degradation.
- This extends balloon life and reduces waste to environmental and economic benefit.

Cover Letter

Most journal manuscript submissions must be accompanied by a cover letter addressed to an editor. As already discussed, it can be advantageous to review the journal editors and select an especially appropriate editor prior to submission if allowed by the journal, otherwise it is generically addressed to the editor-in-chief.

There is some degree of freedom in the layout of this letter, but it should generally be written on university or institution-specific letterhead by the corresponding author. The letter does not need to be long, typically 1-2 pages (maximum including all content and comprising as few as 1 and no more than 3-4 paragraphs of main body text) is sufficient to cover the necessary points. However, some journals have specific requirements for cover letter content so be sure to check the Guide for Authors in advance.

Generally, the most important aspect of a cover letter is a comprehensive and convincing statement of the novelty of the manuscript being submitted and its significance to the scientific community. What is new and why? Why is this important? Ensure that the value and potential impact of the results are strongly emphasised.

In addition to these critical points, cover letters should also generally include:

- Addressee (editor) details and date of submission.
- Title and type of manuscript being submitted.

- Manuscript word, figure, table and reference count (as relevant based on journal requirements).
- Author names and affiliations.
- Contact details for the corresponding author.
- A short overview of the manuscript topic and its relevance to the journal scope.
- Relevant author experience and expertise (for review manuscripts).
- Confirmation that all authors have reviewed and agreed to the submission.
- Confirmation that the manuscript has not been previously published and has not been submitted to another journal concurrently.

Prepare the cover letter in Microsoft Word and save it as a PDF file ready for upload during submission. Alternatively, if a text field is provided in the editorial manuscript submission system for the cover letter, then the prepared letter contents can be pasted into this field as plain text.

Supplementary Material

Supplementary material can be uploaded to support a manuscript submission if desired. More information on supplementary material can be found in the Using this Guide — Journal Articles — Supplementary Material section of this guide.

MANUSCRIPT SUBMISSION

Submission of a manuscript to a journal is usually a fairly fast and straightforward process. Many journals have user friendly online manuscript submission and management platforms, although some smaller publishers still simply request manuscript submissions via email. If a journal only accepts submissions via email, then check their guidelines for document requirements and send the documents to the specified email address as attachments. The contents of the cover letter can be pasted into the email to form its text body.

If an online submission platform exists, then it is usually accessible through the journal's website via a clearly labelled link e.g., 'submit your paper/article'. Registration for a user account in the system is then normally necessary to proceed. Some publishers, such as MDPI, use more generic platforms that allow the creation of a single user account for multiple journals. Authors can then select a journal for submission from a list. However, it is generally necessary to register a separate author account with each journal, even if they are from the same publisher and use the same style online platform. The registration process is the same each time but uses a unique weblink provided on the respective journal's website. A single account for any given journal can usually be used in multiple roles e.g., author, reviewer (and less relevantly as editor or publisher).

Mandatory information required for registration of an author account typically includes:

- Email address (with confirmation process). This can be either personal or institutional.
- Username (if different from the email address) and password.
- Personal information including name and title.
- Institutional information including role, department, institute or school and address.
- Research interests or areas of expertise, which may be used to establish author suitability for reviewer assignment on other manuscripts being considered by the journal.

Although not compulsory, it is often recommended to link journal author profiles with an Open Researcher and Contributor ID (ORCiD) at this point. Many journals provide the opportunity to connect an ORCiD account with a journal author account by providing the ORCiD as text or simply clicking a link and signing in to the ORCiD account. ORCiDs can usually be added or updated in the journal author account settings right up until a manuscript is published. For more information on ORCiDs see Using this Guide — Indexing and Author Profiles — Open Researcher and Contributor ID.

Once a journal author account has been created the submission process can be initiated by logging in and selecting the relevant new submission link. The submission process is typically structured into several pages, the order of which may vary by journal. Once a submission has been started it can be saved and returned to at a later point for completion. However, most journals have a set time frame for draft submissions in their system and will automatically delete them after a predefined period following automated warning emails to the authors. It is subsequently a good idea to ensure that a manuscript is at least close to ready before starting the submission process.

Manuscript Information

Manuscript information is typically requested at the start of the submission process but may sometimes appear after file upload or even at the end of the submission process. The following are examples of information that may be required:

- Article type e.g., original research, review (as relevant for this guide although other article types also exist, see USING THIS GUIDE — JOURNAL ARTICLES — COMMON JOURNAL ARTICLE TYPES). It may also be necessary to indicate if an article belongs to a special issue at this point. Special issues might be e.g., editor invited or be associated with a conference.
- Title (may be automatically checked to ensure that it does not exceed a maximum number of words).
- Manuscript keywords (the number of which may be automatically checked against a limit. They may also have to be chosen from a predefined word pool).
- Abstract (word count may be automatically assessed and checked against the journal's limit).
- Manuscript classification (a set number of themes must be selected from a predefined journal topic hierarchy).
- Authors and affiliations (must often be manually added based on textbox prompts. Institutions may be automatically checked against databases).

During the author addition process there are generally set prompts for selecting author order e.g., numbered boxes, and specifying the corresponding author. If the submitting author is not the corresponding author, which theoretically should not be the case but can in practice vary, then it can be beneficial to postpone selecting another author as the corresponding author until the very end of the submission process. This is because some journals will automatically move the entire submission to the corresponding author's account as soon as this change is made. Once this move has occurred, only the newly selected corresponding author will be able to access the saved submission.

Other common requests during the submission of manuscripts to journals include:

- Manuscript details, such as word, page, figure, or table count.
- Details on funding sources. If a manuscript received no specific financial support, then this should be noted, otherwise provide the relevant details e.g., student scholarship, industry funding, project or grant numbers.
- Acknowledgement of compliance with journal terms and conditions and confirmation that the work being submitted has not been previously published (except as an abstract, published lecture or academic thesis) or submitted elsewhere, has been approved by all authors and will not be republished in the future without prior publisher permission.
- Suggestions of other experts working in the research field who could review the manuscript (typically 3-6) including name, affiliation, email address and justification for the recommendation.

Identifying suitable reviewers for a manuscript submission can be somewhat challenging for those new to the process. If in doubt, the manuscript's reference list can be a good place to start since it contains lists of authors of other scientific studies relevant enough to the manuscript topic to have been cited. Read through these names and investigate the authors' research backgrounds to identify suitable reviewers.

Theoretically, anybody with a scientific background and considerable expertise in the topic of the manuscript, such as other academics or industry specialists, can act as a reviewer. A reviewer should, however, be unbiased and hence have no conflict of interest with the publication of the manuscript i.e., they should not stand to benefit from its publication in any way. This is not limited to those who may stand to gain from the publication financially e.g., work or consult for, own shares in or receive funding from any company or organisation that is linked with the research, but also anyone who may benefit intellectually or socially from the publication e.g., someone who was involved in the project or is working closely with the authors. Some journals may have additional requirements for reviewers e.g., the possession of a Ph.D.

File Upload and Submission

Once all manuscript information has been provided the relevant files can usually be uploaded. Required documents, accepted formats and maximum file sizes are usually provided on the webpage at the file upload stage but if in doubt be sure to refer to the Guide for Authors, which usually contains more detailed information on this topic.

Files are typically uploaded one at a time with a description added, a category selected from a drop-down menu or both. File ordering may be automatically assigned by the online platform based on the category selected during upload or can be manually modified, usually using numbered text boxes.

Manuscripts must typically be uploaded in an editable format e.g., a Microsoft Word document. Requirements for this document vary. It may contain all manuscript content including figures, tables, and captions. Alternatively, it may contain only the text body ('Introduction', 'Materials and Methods', 'Results and Discussion', 'Conclusion') and references with the other content provided in separate documents. These separate documents are detailed during the upload phase or in the Guide for Authors and may include:

- A title page containing the manuscript title, authors, affiliations, abstract and keywords.
- Each figure or table accompanied by its caption as a series of separate Microsoft Word documents.
- Image files for all figures with captions provided in a list at the end of the manuscript.

In addition to the manuscript content, either formatted as a single document or broken into separate documents, the following files must also be typically uploaded during submission:

- A cover letter addressed to a journal editor. Sometimes the cover letter is requested as an uploaded file while in other cases a text field is supplied in the webpage for this content. A text field may also be provided for the recommendation of a handling editor. For more details on cover letters see Publishing Journal Articles — Formatting a Manuscript for Submission — Common Submission Supplements — Cover Letter.
- A declaration of conflicts of interest statement and an author credit statement, see Publishing Journal Articles — Authorship, both potentially included at the end of the manuscript document, as a separate file or in a text field on the webpage.

Some journals will also require the upload of a highlights document, graphical abstract or similar. These should generally be formatted and uploaded as separate documents. See Publishing Journal Articles — Formatting a Manuscript for Submission — Common Submission Supplements — Highlights and Graphical Abstract for more information on these documents.

Options are also usually provided for the upload of supplementary material and linking to stored data or data- or methods-only articles e.g., Data in Brief and MethodsX (see Using this Guide — Journal Articles — Supplementary Material and Publishing Journal Articles — Formatting a Manuscript — Data Availability Requirements).

Once all manuscript information has been provided, and the necessary files uploaded, the author is usually presented with a summary of the submission. This should be checked for accuracy. The submission platform then compiles all the information and files into a single PDF. This PDF will be sent to the reviewers, who will use it to make their assessment, so ensure that it is formatted correctly.

The compiled PDF can sometimes be downloaded and previewed from the submission page. Alternately, it may appear in another section of the online platform, the location of which is specified on the submission page. If the PDF preview contains errors, then it is possible to return to the manuscript information and file upload sections, modify them and then rebuild the PDF. Once the PDF has been thoroughly checked it can be submitted and the manuscript submission process is complete.

Although the reviewers only receive this PDF, all original files are retained in the online platform and can be substituted for updated files over the course of the revision process. It is these individual files and not the PDF that are used in the final assembly of the article by the publisher.

Most online editorial platforms provide authors with updates on the status of their manuscript. Authors must simply log into their journal accounts to check the progress of a submission. This status typically comprises a short description of a few words and can give authors a rough idea of whether their manuscript is with the editor, under review, awaiting a decision, accepted, rejected, sent to production or similar. These status labels vary by platform, are automatically generated based on predefined editor activity and entail broad descriptions of sometimes quite complex internal processes. They are subsequently not necessarily entirely accurate.

PEER REVIEW AND REVISIONS

Manuscript submissions are initially reviewed by an editor to establish if they are suitable for publication in the journal in question. They must be formatted to the journal's requirements and meet its scope. While these are arguably the only requirements that a manuscript must meet to be considered for publication in a journal, editors also often consider the manuscript's novelty, potential impact, appeal to the journal's readership and ability to promote the journal's interests and reputation during this initial editorial assessment.

Desk Rejection by the Editor

Initial editorial assessments can be somewhat fickle processes. The volume of submissions that editors receive is far greater than they could possibly publish, and preliminary evaluations can subsequently be quite superficial. Manuscripts that are not formatted to the journal requirements are often instantly rejected. Hence the emphasis on this step. Even if all formatting requirements are met, editors may potentially only read the abstract and conclusion of a manuscript and perhaps glance at the figures, tables and timeliness of the references before reaching a decision on whether a manuscript should be rejected immediately or progress to peer review.

Besides incorrect formatting, another common reason for manuscript desk rejection is a lack of compliance with scope. Journal scope is often quite vaguely defined using broad keywords and themes and occasionally explicit examples of inclusions or exclusions. Interpretation of these definitions in any given case is unfortunately often based on the personal opinion of the editor and can vary radically even within the same journal. It is not uncommon for manuscripts that would appear to fit a journal's scope to be rejected on scope-based grounds while in other cases journals end up publishing manuscripts that don't really seem to fit their scope at all. Be prepared for rejection without any apparent rhyme or reason at an early stage in the publishing process. Such rejections are not necessarily attributable to poor manuscript quality or incompetent journal selection but rather the individual tastes of editorial staff.

The Peer Review Process

If an editor judges a manuscript to be suitable for publication in a journal, then they will send it out for peer review. The peer review process is an independent assessment of the manuscript's quality by other experts in the field and is crucial to maintaining a high level of scientific quality and integrity in the literature.

Reviewer Assignment

Ultimately, reviewer selection is completely at the discretion of the editor managing the manuscript submission. Some journals have pools of trusted reviewers who they utilise preferentially. However, they will often use author-supplied suggestions when assigning reviewers (see Publishing Journal Articles — Manuscript Submission — Manuscript Information). The identity of the reviewers is not usually revealed to the manuscript authors, although the identity of the authors is known to the reviewers. This reviewer anonymity is maintained to encourage uninhibited and honest feedback on manuscripts under consideration. A 'double blind peer review' where the identity of the authors is also withheld from the reviewers can be requested during the submission process if desired. Authors can also request that specific individuals be excluded from the reviewer pool during manuscript submission although there should generally be a reason for doing so.

The suitability of a reviewer beyond the recommendation of the authors and the editor's opinion is entirely self-moderated. The editor will send the potential reviewer an email with the title and abstract of the manuscript inviting them to review it and it is then up to the reviewer to decide if they are suitably qualified and interested in doing so.

Peer Review Timelines

Reviewing a scientific manuscript is not paid or compensated in any way, except for some journals which offer credits towards publishing costs in exchange for review work. It is rather a service to the scientific community that most academics will oblige and take seriously. Nonetheless, scientists often have many demands on their time and review work, while important, is not often a top priority. It is subsequently ill-advised to select especially prominent members of the scientific community, such as renowned professors, as reviewers. Through no fault of their own, many simply won't be able to review manuscripts in a timely manner, if at all.

Journals try to expedite the peer review process by imposing deadlines on reviewers, but even so there is understandably a great deal of respect for the voluntary contribution reviewers make and many journals are very flexible with review timelines. The return of a single review from a reviewer can take weeks or even months and, with several reviewers always used in parallel, receiving all reviews back for a single round of revisions often requires a few months of patience.

Type and Scope of Reviewer Feedback

Reviewers typically assess all aspects of a manuscript including adherence to journal scope and formatting requirements, language and expression, quality of data presentation including figures and tables, novelty and impact of the results, suitability of methodologies used in experimental work, presence of factual or other errors and substantiation of claims. They will typically provide comments (both to the authors and confidentially to the editor) and an indication of whether the manuscript can be accepted in its current form or if it requires minor or major revisions. Many journals also require reviewers to assign a numerical grade to the quality of the manuscript and answer a set of questions relating to the suitability of the manuscript for publication in the journal.

The type and extent of feedback provided by a reviewer will vary from person to person. Some reviewers may home in on specific aspects of a manuscript with which they are especially experienced, such as statistics or particular experimental methodologies; others may provide more general feedback. It is not uncommon to have vastly different assessments of quality from separate reviewers for the same manuscript based on their own individual interpretations, varied skills, and degrees of experience. It is for this reason that manuscripts are normally reviewed by at least 2-3 people. That said even more thorough reviews (4-6 reviewers) are not uncommon, especially if there is a lack of consensus between reviewer reports during the first round of the peer review process.

Editorial Decision – Acceptance, Rejection or Revision

Once the editor is satisfied with the quality of the review, they will compile the reviewer reports and their own editorial suggestions into a response letter to the authors. In this letter they will outline if the submitted manuscript has been accepted, rejected, or requires minor or major revisions. Manuscripts are rarely accepted for publication in a journal without revisions but if this is the case then no further action is required from the authors.

If the manuscript was rejected following peer review, then it is unlikely to be on the grounds of scope but rather due to problems with the manuscript. Problems associated with a rejection rather than an invitation to resubmit with major revisions may be considerable. Nonetheless, the reviewer feedback provided alongside the rejection can be very useful in improving manuscripts and increasing the chances of acceptance for the next submission. Resubmission of a rejected manuscript should generally be to a journal other than that to which it was originally submitted unless such a resubmission was explicitly suggested by the editor at the time of the rejection.

Manuscript Revisions and Resubmission

If minor or major revisions are requested, then authors are expected to carefully consider the reviewers' feedback and make appropriate changes to their manuscript before resubmission. Minor and major are somewhat subjective terms in this context and different reviewers and editors have different ideas of what these descriptions mean. Minor revisions could be errors in the manuscript text, figures or tables that need to be corrected while major revisions may be text-based but could also indicate fundamental errors, oversights or omissions in conceptual understanding, experimental methodology, interpretation of results or similar.

Authors must provide a response to every comment made by the editor and reviewers or risk rejection upon resubmission. As time consuming as addressing reviewer comments can be, insight and perspective from other experts in the field can radically improve a manuscript. Authors should remember that reviewers' comments are aimed at helping them and safeguarding the quality and integrity of the research area. Try to be as open as possible to criticism. Consider every comment carefully and if possible, implement it. Reviewers often have valuable suggestions, feedback, and perspective.

Response to Reviewers' Comments

A response to the reviewers' comments detailing changes made in the revised manuscript should be compiled in a separate document. This document has no special formatting requirements. It can be formulated on university letterhead if desired but for many is neatly arranged yet fairly plain. To prepare this response paste all remarks, queries, and suggested changes from the reviewers (and optionally also from the editor) into a Microsoft Word document. Break every individual entry from each person into a separate paragraph so that they can be addressed point-by-point. Every thematically different element of each remark, query or suggestion should be separated to ensure a clear and focussed response can be made to every aspect of every entry.

It is not compulsory to make every change that a reviewer has requested. If there is a good reason to justify not making a requested change e.g., the reviewer has misunderstood an element of the manuscript or has touched on a controversial topic, then it is acceptable to rebut the comment. However, if a requested revision cannot or should not be made, then at least attempt to clarify the potential for misunderstandings in the manuscript text and reinforce the original statement with additional context, relevant experimental work, or references. It should be at least apparent that the comment has been considered at length and every attempt has been made to address it in the most appropriate way. In general aim to implement all reviewer feedback unless there is a good reason not to and, if so, then be sure to state and discuss this reason in detail in the response to reviewers' comments document.

Structure the response document using appropriate labels such as:

- 'Reviewer comments' arranged by reviewer number and broken into individually labelled 'reviewer queries' which are then addressed with an 'author's response'.
- A detailed 'author's response' to each reviewer query that comprehensively describes the changes made in the revised manuscript based on the query or provides additional information or justification as to why no change has been made.
- 'References' to support discussion or rebuttal of reviewer queries. If a reviewer requested change has not been implemented, then a justification backed by references is advisable. If additional references were explicitly requested by the reviewers for any reason, then they should also be included.

Revised Manuscript

Revisions should be made in a revised manuscript document. Create a new copy of the original manuscript and retitle it to reflect its revised nature. Work through the comments one at a time, making the appropriate changes in the revised manuscript, marking them using the journal's convention, which is often coloured or highlighted text, and then providing a detailed description of the changes made in the response to reviewers' comments document. Journals do not usually accept the use of track changes or other review tools in Microsoft Word for revisions, although some exceptions do exist e.g., MDPI journals. Implement revisions in the manuscript body as plain text instead.

Resubmission Cover Letter

Finally, a new cover letter must be formulated to accompany the revised manuscript and response to reviewers' comments document. This cover letter is addressed to the editor so should contain a point-by-point response to any editorial queries or concerns that the editor raised during the initial submission process and comment on the changes made to the manuscript based on the reviewers' comments e.g., how comprehensively they were able to be addressed. The cover letter generated in the PUBLISHING JOURNAL ARTICLES — COMMON SUBMISSION SUPPLEMENTS — COVER LETTER section of this guide can be used as a template (at least in terms of letter formatting) for this response but the body text should be replaced. If there are no editorial queries to address, or they are instead included in the response to reviewers' comments document then the cover letter can be quite short. Simply outline the attachments, comment on the revisions made and any conventions used while revising the manuscript.

Resubmission

The final files can then be resubmitted to the journal using the electronic manuscript management system. Manuscripts that were assigned minor revisions during peer review may be accepted by the editor upon resubmission without re-review. Manuscripts that required major revisions are often sent again for peer review. Reviewers who requested major revisions to the initial submission are normally expected to check the manuscript a second time to ensure that it has been revised to their satisfaction, but the editor may also invite additional new reviewers to assess the revised manuscript at this stage. Most manuscripts will require between 1-3 rounds of revision prior to acceptance.

Article Production and Correction of Proofs

Once a manuscript has been accepted for publication, the editor will contact the authors with the news and send the manuscript to the publisher's editorial staff for production. There it will normally be copy edited, the references will be checked, and it will be formatted to the journal style. Some publishers will also provide options for graphic design support or development of additional interactive multimedia at this stage to accompany the publication on their website.

A 'proof' of the final version of the article will be sent to the authors for their approval prior to publication. This is sometimes in the form of a PDF with accompanying instructions on how to request changes but is often through an online platform where text can be directly modified, comments made, and a sample PDF generated. Queries from publisher-based editorial staff, often consisting of name checks and other generic administrative questions, must also be addressed at this stage. Check proofs carefully. This is the last opportunity to request any changes that need to be made prior to final publication.

The journal's scientific editors and publisher-based editorial staff are quite distinct. During the initial submission and peer review process, the manuscript submission is handled by scientific editors, who are typically academics employed by universities that serve the journal in an honorary or sometimes paid capacity. However, as soon as a manuscript is accepted and sent to the publisher for production, it will be managed by more generic editorial staff. It is subsequently important to check the final copy-edited proof to ensure that scientific meaning has not been altered during this process. It is also worth noting that the journal's scientific editorial staff have little influence beyond the point of manuscript acceptance and subsequently cannot do much if mistakes are made during the proofing or production processes. Post-publication mistakes must typically be fixed by issuing an erratum or corrigendum (see Using this Guide — Journal Articles — Common Journal Article Types). This is an additional document published as a supplement to the article and can be arranged by contacting the journal's scientific staff.

Final Paperwork

The completion of the publishing process is accompanied by paperwork signed by or on behalf of the corresponding author. This may include:

- Assignment of the copyright and full redistribution and reuse rights for the manuscript, its supplementary material and research data to the publisher.
- Confirmation that all content being transferred belongs to the authors, is original, does not infringe or violate any copyright or intellectual property laws and meets all ethical and sanctions requirements.
- Acknowledgement that the content cannot be revised or modified without the publisher's consent.
- Description of the author's usage rights for the published content (be it the pre-print, accepted manuscript, or published journal article) for personal and internal institutional use or scholarly sharing.
- Options for subscription-based or open access licensing. If the article will be available through subscription only, then there is usually no cost pertaining to the publication applicable to the authors. If there is a cost, this could be an indication of predatory behaviour by the journal (see USING THIS GUIDE — JOURNAL ARTICLES — JOURNAL REPUTATION AND METRICS). If a journal offers open access options and such options are selected, then a fee will be billable to the authors. Some research institutions cover this fee on behalf of their staff (see USING THIS GUIDE — JOURNAL ARTICLES — JOURNAL ARTICLE ACCESS AND LICENSING).

AUTHOR PROFILE

Mitchell P. Jones is an academic, professional engineer and chartered chemist. He obtained doctorates in chemistry and mechanical and manufacturing engineering from the University of Vienna, Austria, and The Royal Melbourne Institute of Technology University, Australia. Mitchell has worked at leading universities, defence and aerospace companies spanning Australia and Europe and has an expansive publication record in top materials science and biotechnology conferences and journals. His research has received international attention from news media including *The New York Times*, *Daily Mail*, *CNN London*, *CBC*, *Materials Today*, *Der Standard*, *Der Spiegel* and *Neue Zürcher Zeitung*. He also currently serves on the i-SMaRT conference advisory committee, the Environmental College Board of The Institution of Engineers Australia and is the Outreach Editor for the journal *Resources, Conservation and Recycling*.